After Method

CW00419579

'Research Methods': a compulsory course, which is loved by some but hated by many! This stimulating book is about what went wrong with 'research methods'. Its controversial argument is radical, even revolutionary.

John Law argues that methods don't just describe social realities but also help to create them. The implications of this argument are highly significant. If this is the case, methods are always political, and this raises the question of what kinds of social realities we want to create.

Most current methods look for clarity and precision. It is usually said that messy findings are a product of poor research. The idea that things in the world might be fluid, elusive, or multiple is unthinkable. Law's startling argument is that this is wrong and it is time for a new approach. Many realities, he says, are vague and ephemeral. If methods want to know and to help shape the world, then they need to reinvent their practice and their politics in order to deal with mess. That is the challenge. Nothing else will do.

This book is essential reading for students, postgraduates and researchers with an interest in methodology.

John Law is Professor of Sociology and Technology Studies at Lancaster University. He has written widely on social theory, methodology, technologies, and health care.

International Library of Sociology
Founded by Karl Mannheim
Editor: John Urry, *Lancaster University*

After Method
Mess in social science research

John Law

Routledge
Taylor & Francis Group

LONDON AND NEW YORK

First published 2004
by Routledge
2 Park Square, Milton Park, Abingdon, Oxon OX14 4RN

Simultaneously published in the USA and Canada
by Routledge
270 Madison Ave, New York, NY 10016

Routledge is an imprint of the Taylor & Francis Group

© 2004 John Law

Typeset in Garamond by Keystroke, Jacaranda Lodge, Wolverhampton
Printed and bound in Great Britain by Cromwell Press, Trowbridge,
Wiltshire

All rights reserved. No part of this book may be reprinted or
reproduced or utilised in any form or by any electronic, mechanical,
or other means, now known or hereafter invented, including
photocopying and recording, or in any information storage or
retrieval system, without permission in writing from the publishers.

British Library Cataloguing in Publication Data
A catalogue record for this book is available from the British Library

Library of Congress Cataloging in Publication Data
A catalog record for this book has been requested

ISBN 0–415–34174–4 (hbk)
ISBN 0–415–34175–2 (pbk)

Contents

Acknowledgements

This book grows out of the writing, the conversation, the friendship, and the support of a large number of colleagues, friends and students, a context that has grown and been sustained over many years. Amongst these people I would like in particular to thank: Madeleine Akrich; Kristin Asdal; Andrew Barry; Ruth Benschop; Brita Brenna; Michel Callon; Claudia Castañeda; Bob Cooper; Anni Dugdale; the late Edith Eldridge; Donna Haraway; Hans Harbers; Dixi Henriksen; John Holm; Casper Jensen; Torben Jensen; Karin Knorr-Cetina; Bruno Latour; Maureen McNeil; Turid Markussen; Ivan da Costa Marques; Tiago Moreira; Bernike Pasveer; Jeannette Pols; Vololona Rabeharisoa; Lars Risan; John Staudenmaier sj; Marilyn Strathern; Lucy Suchman; Nigel Thrift; David Turnbull; John Urry; Marja Vehvilainen; Laura Watts; and Steve Woolgar. In one way or another, in person or through their writing, all these people have inspired my interest in the topic of method, and have helped to shape the arguments in the book. A number of them have read it in earlier drafts and offered extensive comments. I am most grateful to them all.

In addition to this larger group, five friends and colleagues have been particularly important in helping to give the book its shape and form, in sustaining my efforts as I have attempted to clarify its arguments. I am therefore deeply grateful: to Kevin Hetherington, whose shared concern with the indirections of allegory is central to this book, and whose conversations, usually over the supper-table, have been a source of continuing support and insight, to Annemarie Mol, who invented difference and multiplicity, and who, often in the course of energetic walks, has debated, encouraged, inspired and resisted the extension of those arguments in their present form at every stage; to Ingunn Moser, whose interest in complex subjectivities, embodiments, distributions and the elusive, often discussed in the course of even more energetic walks, has been the occasion for exploring many of the positions argued in our joint work and in this book; to Vicky Singleton, who fortunately walks somewhat more slowly, but whose sensibility to and work on the elusive, the hidden, the muchness of the world, and things that don't quite fit, has deeply informed both our collaborative writing and the arguments as they are developed here; and to Helen Verran, who also walks more slowly, but whose work on ontic/epistemic imaginaries nevertheless travels long distances, and whose generous conversations have been crucial, both for clarifying many specificities about Aboriginal history and practice, and more generally as an inspiration in thinking about method, realities, and their possibilities. So though words cannot fully catch their contributions, I thank these five friends in particular. Of course, I am responsible for the form their arguments take in this text.

I am also most grateful to Sheila Halsall, Angus Law and Duncan Law. The book would not have been possible without their continued personal support and intellectual encouragement. It has also been a particular pleasure to debate many of its arguments with Duncan at every stage of their development, and I am most grateful to Sheila Halsall for her photographic inspiration.

Finally, I am grateful to the Centre for Science Studies, the Department of Sociology, and the Faculty of Social Science, all at Lancaster University. Lancaster is a creative and supportive intellectual environment for social science inquiry, and as a part of this it generously offers sabbatical leave to its faculty. The first draft of this book was written during a period of such leave between September and December, 2001.

Note

Questions of method arising from this book are debated at the Lancaster University Sociology department website. Please visit http://www.comp.lancs. ac.uk/sociology/

1 After method: an introduction

If this is an awful mess . . . then would something less messy make a mess of describing it?

> 'There is no use in trying,' said Alice; 'one can't believe impossible things.' 'I dare say you haven't had much practice,' said the Queen. 'When I was your age, I always did it for half an hour a day. Why, sometimes I've believed as many as six impossible things before breakfast.'
>
> (Lewis Carroll, *Alice in Wonderland*)

How might method deal with mess?

Look at the picture above, and at the question posed by the caption. This book is about that caption, and about what happens when social science tries to describe things that are complex, diffuse and messy. The answer, I will argue, is that it tends to make a mess of it. This is because simple clear descriptions don't work if what they are describing is not itself very coherent. The very attempt to be clear simply increases the mess. So the book is an attempt to imagine what it might be to remake social science in ways better equipped to deal with mess, confusion and relative disorder.

No doubt some things in the world can indeed be made clear and definite. Income distributions, global CO_2 emissions, the boundaries of nation states, and terms of trade, these are the kinds of provisionally stable realities that social and natural science deal with more or less effectively. But alongside such phenomena the world is also textured in quite different ways. My argument is that academic methods of inquiry don't really catch these. So what are the textures they are missing out on?

If we start to make a list then it quickly becomes clear that it is potentially endless. Pains and pleasures, hopes and horrors, intuitions and apprehensions, losses and redemptions, mundanities and visions, angels and demons, things that slip and slide, or appear and disappear, change shape or don't have much form at all, unpredictabilities, these are just a few of the phenomena that are hardly caught by social science methods. It may be, of course, that they don't belong to social science at all. But perhaps they do, or partly do, or should do. That, at any rate, is what I want to suggest. Parts of the world are caught in our ethnographies, our histories and our statistics. But other parts are not, or if they are then this is because they have been distorted into clarity. This is the problem I try to tackle. If much of the world is vague, diffuse or unspecific, slippery, emotional, ephemeral, elusive or indistinct, changes like a kaleidoscope, or doesn't really have much of a pattern at all, then where does this leave social science? How might we catch some of the realities we are currently missing? Can we know them well? *Should* we know them? Is 'knowing' the metaphor that we need? And if it isn't, then how might we relate to them? These are the issues that I open up in this book.

I don't have a single response to these questions. The book is intended as an opening rather than a closing. In any case, if much of reality is ephemeral and elusive, then we cannot expect single answers. If the world is complex and messy, then at least some of the time we're going to have to give up on simplicities. But one thing is sure: if we want to think about the messes of reality at all then we're going to have to teach ourselves to think, to practise, to relate, and to know in new ways. We will need to teach ourselves to know some of the realities of the world using methods unusual to or unknown in social science.

For example? Here are some possibilities. Perhaps we will need to know them through the hungers, tastes, discomforts, or pains of our bodies. These would be forms of knowing as embodiment. Perhaps we will need to know them through 'private' emotions that open us to worlds of sensibilities, passions, intuitions, fears and betrayals. These would be forms of knowing as emotionality or apprehension. Perhaps we will need to rethink our ideas about clarity and rigour, and find ways of knowing the indistinct and the slippery without trying to grasp and hold them tight. Here knowing would become possible through techniques of deliberate imprecision. Perhaps we will need to rethink how far whatever it is that we know travels and whether it still makes sense in other locations, and if so how. This would be knowing as situated inquiry. Almost certainly we will need to think hard about our relations with whatever it is we know, and ask how far the process of knowing it also brings it into being. And as a theme that runs through everything, we should certainly be asking ourselves whether 'knowing' is the metaphor that we need. Whether, or when. Perhaps the academy needs to think of other metaphors for its activities – or imagine other activities.

Such talk is new but at the same time it is not so new. There are many straws in the social science wind which suggest that it is starting to blow in directions such as these. Over the last two decades methods for the analysis of visual materials, performance approaches, and an understanding of methods as poetics or interventionary narrative have all become important. Students of anthropology, cultural studies and sociology have grappled with ways of thinking about and describing decentred subjectivities and the geographical complexities that arise when intimacy no longer necessarily implies proximity. There is also a developing sense that global flows are uncertain, unpredictable indeed chaotic in the mathematical sense. Many now think that ethnography needs to work differently if it is to understand a networked or fluid world. The sense that knowledge is contexted and limited has become widespread, and feminists have talked of situated knowledges while anthropologists have explored writing and receiving culture.[1] Market research, often more imaginative than academic social science, has developed methods such as tasting panels for understanding the non-cognitive and the ephemeral. And never to be outdone, management consultancy has adopted 'soft methods' for intervening in organisations by turning to dramatisations, enactments and performances.

So the world is on the move and social science more or less reluctantly follows. Agency is imagined as emotive and embodied, rather than as cognitive: the nature of the person is shifting in social theory and practice. Structures are imagined to be more broken or unpredictable in their fluidity. But at the same time, within social science, talk of 'method' still tends to summon up a relatively limited repertoire of responses. The collection and manipulation of certain kinds of quantitative data is emblematic for research methods in many parts of social science including much of sociology, economics, psychology, and human geography. The collection and manipulation of certain kinds of qualitative materials is iconographic in anthropology, cultural studies, science studies, and other parts of sociology and human geography. The quantitative/qualitative iconography – and its division – is built into many courses on research methods. In the English-speaking world it is unusual, perhaps impossible, to qualify as a degree-level social scientist without following such courses and learning the appropriate suites of methods. Indeed, national recognition of social science courses in the United Kingdom now demands that these include both quantitative and qualitative methods, though many students and teachers dislike such courses and find their content to be at best marginally relevant to the research process.

This book makes a sustained argument for a way of thinking about method that is broader, looser, more generous, and in certain respects quite different to that of many of the conventional understandings. It is therefore, in part, an attack on the limits set by such understandings. But there are various reasons why any such attack needs to be cautious. One is that 'social science method' is an encouragingly multi-headed beast. It is already variegated and hetero-geneous in its claims, but even more so in its practices. Since I am arguing for greater methodological variety, existing variety is surely welcome wherever it is to be found – which is everywhere. This suggests, then, that the problem is not so much lack of variety in the *practice* of method, as the hegemonic and dominatory pretensions of certain versions or *accounts* of method. I will return to this question, that of the normativity of method, shortly.

Another reason for caution is that standard research methods are often important, not to say necessary. To take one notorious example, it was quantitative epidemiological research that established a plausible link between smoking and lung cancer.[2] Another example with a more social science flavour would be the many studies, again often quantitative, that have revealed strong relations between poor health and a range of social inequalities including poverty.[3] Or between vulnerability to disaster, and age, social isolation and poverty.[4] There are, to be sure, always complexities and ambivalences.[5] Never-theless studies such as these have been the basis for major health education campaigns. And endless other success stories for standard methods, quantitative and qualitative, could be cited.

It cannot be the case, then, that standard research methods are straight-forwardly wrong. They are significant, and they will properly remain so. This is why I say that I am after a broader or more generous sense of method, as well as one that is different. But to talk of difference is indeed to edge towards criticism. As I have suggested above, I want to argue that while standard methods are often extremely good at what they do, they are badly adapted to the study of the ephemeral, the indefinite and the irregular. As I have just suggested, this implies that the problem is not so much the standard research methods themselves, but the normativities that are attached to them in discourses about method. If 'research methods' are allowed to claim methodological hegemony or (even worse) monopoly, and I think that there are locations where they try to do this, then when we are put into relation with such methods we are being placed, however rebelliously, in a set of constraining normative blinkers. We are being told how we must see and what we must do when we investigate. And the rules imposed on us carry, we need to note, a set of contingent and historically specific Euro-American assumptions.[6]

Here the problem is not that our research methods (and claims about proper method) have been constructed in a specific historical context. *Everything* is constructed in a specific historical context and there can be no escape from history. Rather it is that they, or at least their advocates, tend to make excessively general claims about their status. The form of argument is often like this (think, for instance, about rules for statistical sampling, or avoiding leading questions in interviews). 'If you want to understand reality properly then you need to follow the methodological rules. Reality imposes those rules on us. If we fail to follow them then we will end up with substandard knowledge, knowledge that is distorted or does not represent what it purportedly describes.' There are two things I want to say in response to such suggestions about the importance of methodological rule-following. The first is counter-intuitive. It is that methods, their rules, and even more methods' practices, not only describe but also help to *produce* the reality that they understand. I will carefully explore the reasons for making this suggestion in due course. However, for the moment let me simply note that there is a fair amount of heavyweight work on the history of science and social science that makes precisely this argument. Perhaps again counter-intuitively, I will also say that if methods tend to produce the reality they describe, then this may be, but is not necessarily, obnoxious. Again I will return to this argument at some length in due course. But what is important now is to note that if these two claims are right then they have profound implications for our understanding of the nature of research.

There is a further and more straightforward point to be made. This is that claims about the general importance of methodological rules also tend to get naturalised in social science debate. *Particular* sets of rules and procedures may be questioned and debated, but the overall need for proper rules and procedures is not. It is taken for granted that these are necessary. And behind the assumption that we need such rules and procedures lies a further range of assumptions that are also naturalised and more or less hidden. These have to do with what is most important in the world, the kinds of facts we need to gather, and the appropriate techniques for gathering and theorising data. All of these, too, are naturalised in the common sense of research. Yes, things are on the move. Nevertheless, the 'research methods' passed down to us after a century of social science tend to work on the assumption that the world is properly to be understood as *a set of fairly specific, determinate, and more or less identifiable processes.*

Within social science conventions, which are the best methods (and theories) for exploring those somewhat specific processes? This is a matter for endless debate. Neo-Marxists discover world systems, or uneven developments, or they theorise regulation. Foucauldians discover systems of governmentality. Communitarians discover communities and the need for informal association and responsibility. Feminists discover glass ceilings, cultural sexisms, or gendering assumptions built into scientific and social science method. As a part of this, social science common sense also assumes that society changes. Indeed this is one of the rationales for social science: that it can participate in and guide that change. (Witness the health-inequalities finding mentioned above, but also the larger political inheritance of Euro-American social theory.) But, overall, the social is taken to be fairly definite. Such is the framing assumption: that there are definite processes out there that are waiting to be discovered. Arguments and debates about the character of social reality take place *within* this arena. And this is what social science is meant to do: to discover the most important of those definite processes. But this is precisely the problem: *this is not necessarily right*. Accordingly, it indexes the broadening shift that I want to make. The task is to imagine methods when they no longer seek the definite, the repeatable, the more or less stable. When they no longer assume that this is what they are after.

So what are those elusive realities? This is for discussion. I have my own sense of what it is that might be important and this informs my argument. However, I do not want to legislate a particular suite of research methods. To do so would be to recommend an alternative set of blinkers. Instead I argue that the kaleidoscope of impressions and textures I mention above reflects and refracts a world that in important ways cannot be fully understood as a specific set of determinate processes. This is the crucial point: what is important in the world including its structures is not simply technically complex. That is, events and processes are not simply complex in the sense that they are technically difficult to grasp (though this is certainly often the case). Rather, they are also complex because they *necessarily exceed our capacity to know them*. No doubt local structures can be identified, but, or so I want to argue, the world in general defies any attempt at overall orderly accounting. The world is not to be understood in general by adopting a methodological version of auditing.[7] Regularities and standardisations are incredibly powerful tools but they set limits. Indeed, that is a part of their (double-edged) power. And they set even firmer limits when they try to orchestrate themselves hegemonically into purported coherence.

The need, then, is for heterogeneity and variation. It is about following Lewis Carroll's queen and cultivating and playing with the capacity to think six impossible things before breakfast. And, as a part of this, it is about creating metaphors and images for what is impossible or barely possible, unthinkable or almost unthinkable. Slippery, indistinct, elusive, complex, diffuse, messy, textured, vague, unspecific, confused, disordered, emotional, painful, pleasurable, hopeful, horrific, lost, redeemed, visionary, angelic, demonic, mundane, intuitive, sliding and unpredictable, these are some of the metaphors I have used above. Each is a way of trying to open space for the indefinite. Each is a way of apprehending or appreciating displacement. Each is a possible image of the world, of our experience of the world, and indeed of ourselves. But so too is their combination. What this might mean in practice will be explored below. But together they are a way of pointing to and articulating a sense of the world as an unformed but generative flux of forces and relations that work to produce particular realities.

The world as a 'generative flux' that *produces* realities? What does this mean? I can only tackle this question bit by bit, and any answer will be incomplete. Nevertheless, in this way of thinking the world is not a structure, something we can map with our social science charts. We might think of it, instead, as a maelstrom or a tide-rip. Imagine that it is filled with currents, eddies, flows, vortices, unpredictable changes, storms, and with moments of lull and calm. Sometimes and in some locations we can indeed make a chart of what is happening round about us. Sometimes our charting helps to produce momentary stability. Certainly there are moments when a chart is useful, when it works, when it helps to make something worthwhile: statistics on health inequalities. But a great deal of the time this is close to impossible, at least if we stick to the conventions of social science mapping. Such is the task of the book: to begin to imagine what research methods might be if they were adapted to a world that included and knew itself as tide, flux, and general unpredictability.

This will take us, and uncertainly, far from a conventional discussion of method, but also from our common-sense assumptions about the character of the world and how we come to know it. And this in turn means that it is also important to avoid some possible misunderstandings:

- First, as I have tried to insist above, I am not saying that there is no room for conventional research methods. Such is not at all the point of my argument.
- Second, and more generally, I am not saying that there is no point in studying the world. I am not recommending defeatism. On the contrary, the task is to reaffirm a reshaped set of commitments to empirical and theoretical inquiry. The issue is: what might social science inquiry look like in a world that is an unformed but generative producer of realities? What shapes might we imagine for social science inquiry? And, importantly, what might responsibility be in such a world?
- Third, I am not recommending political quietism. I shall have a lot more to say about politics below, but the basic point is simple. Since social (and natural) science investigations interfere with the world, in one way or another they always make a difference, politically and otherwise. Things change as a result. The issue, then, is not to seek disengagement but rather with how to engage. It is about how to make good differences in circumstances where reality is both unknowable and generative.
- Finally, what I am arguing is not a version of philosophical idealism. I am not saying that since the world defies any overall attempt to describe and understand it, we can therefore realistically believe anything about it we like. I also discuss this much more fully below, but everything I argue assumes that there is a world out there and that knowledge and our other activities need to respond to its 'out-thereness'. It is a world, as I've

suggested, that is complex and generative. I will argue that we and our methods help to generate it. But the bottom line is very simple: believing something is never enough to make it true.

As is obvious, this argument strays into philosophy. Like others working in the discipline of science, technology and society (STS) I have explored how science is practised in laboratories, and it is difficult to do this without tripping over the writings of philosophers of science and social science. Again, like many others in STS, I do not share many of the most widespread philosophical and common-sense understandings about the nature of scientific (and social science) inquiry. To a first approximation, STS argues that science is a set of practices that are shaped by their historical, organisational and social context. It further argues that scientific knowledge is something that is constructed within those practices.[8] Thus though they draw on history and philosophy of science, these kinds of arguments also tread on a lot of philosophical toes. But here we need a health warning. Just as this is not a book on method, conventionally understood, neither is it a text in philosophy of science or social science, conventionally understood. The proof of new ways of thinking about method, or so I take it, lies in their results and their outcomes, rather than in their antecedents. Nevertheless, the arguments that I develop indeed have philosophical antecedents. They draw on parts of the philosophy of science but also on philosophical romanticism and (what is perhaps its contemporary expression) post-structuralism. A few words on these two traditions.

Social science has struggled with the inheritance of philosophical romanticism for at least 200 years (at the same time wrestling with its mirror image, the classical commitment to reason and inquiry, embedded in the Enlightenment project (Gouldner 1973)). I will touch on a few of the relevant arguments later. For now it is simply useful to note that many notable social theorists (to name but a few, Karl Marx, Georg Simmel, Max Weber, Georg Lukács, George Herbert Mead and Walter Benjamin) incorporated important elements of philosophical romanticism in their accounts of the world. This means that in different ways they responded to the idea that the world is so rich that our theories about it will always fail to catch more than a part of it; that there is therefore a range of possible theories about a range of possible processes; that those theories and processes are probably irreducible to one another; and, finally, that we cannot step outside the world to obtain an overall 'view from nowhere' which pastes all the theory and processes together.

A related set of intuitions informs such post-structuralist writers as Michel Foucault, Gilles Deleuze and Jacques Derrida. Instead of assuming that there is a specific external reality upon which we can ground our efforts to know the world, such writers mobilise metaphors such as flux to index the sense that whatever there is in the world cannot be properly or finally caught in the webs of inquiry found in science and social science (or indeed any other form of knowing). And then they talk of 'discourse', 'deferral' or 'episteme' to point to the methodological efforts to make and know limited moments in the fluxes that make up reality.

Philosophical romanticism and post-structuralism have informed some versions of social science (and especially qualitative) method. They have inspired various empirically grounded styles of investigation in sociology, anthropology, cultural studies, feminism, human geography, and science, technology and society (STS). It is, for instance, possible to go to *verstehende* sociology, symbolic interactionism, to anthropology and cultural studies of difference, to post-colonialism, to actor-network theory, or to parts of feminist technoscience studies to see what these intuitions might mean in methodological practice.[9] But even so, as I have noted, more often social (and still more natural) science 'method talk' connotes something quite different – that is a particular version of rigour. This is the idea that it is important to obtain the best and technically robust possible account of reality, where reality is assumed, as I have suggested above, to be a pretty determinate set of discoverable entities and processes. That such is what the world *is*: a set of possibly discoverable processes.[10]

My aim is thus to broaden method, to subvert it, but also to remake it. I would like to divest concern with method of its inheritance of hygiene. I want to move from the moralist idea that if only you do your methods properly you will lead a healthy research life – the idea that you will discover specific truths about which all reasonable people can at least temporarily agree. I want to divest it of what I will call 'singularity': the idea that indeed there are definite and limited sets of processes, single sets of processes, to be discovered if only you lead a healthy research life. I also want to divest it of a commitment to a particular version of politics: the idea that unless you attend to certain more or less determinate phenomena (class, gender or ethnicity would be examples), then your work has no political relevance. I want to subvert method by helping to remake methods: that are not moralist; that imagine and participate in politics and other forms of the good in novel and creative ways; and that start to do this by escaping the postulate of singularity, and responding creatively to a world that is taken to be composed of an excess of generative forces and relations.

To do this we will need to unmake many of our methodological habits, including: the desire for certainty; the expectation that we can usually arrive at more or less stable conclusions about the way things really are; the belief that as social scientists we have special insights that allow us to see further than others into certain parts of social reality; and the expectations of generality that are wrapped up in what is often called 'universalism'. But, first of all we need to unmake our desire and expectation for security.

Method, as we usually imagine it, is a system for offering more or less bankable guarantees. It hopes to guide us more or less quickly and securely to our destination, a destination that is taken to be knowledge about the processes at work in a single world. It hopes to limit the risks that we entertain along the way. Method, then, may allow us to learn that particular hypotheses are wrong: this is an important part of methodology's self-presentation, and it has important merits.[11] It may also allow us to learn that *particular methods* are flawed. But as a framework, method *itself* is taken to be at least provisionally secure. The implication is that method hopes to act as a set of short-circuits that link us in the best possible way with reality, and allow us to return more or less quickly from that reality to our place of study with findings that are reasonably secure, at least for the time being.[12] But this, most of all, is what we need to unlearn. Method, in the reincarnation that I am proposing, will often be slow and uncertain. A risky and troubling process, it will take time and effort to make realities and hold them steady for a moment against a background of flux and indeterminacy.

There is a beautiful book by David Appelbaum called *The Stop* (1995). This contrasts the quickness of seeing with the groping of the blind person. It seems to us, he says, that the blind person lacks vision. No doubt this is right. But Appelbaum's argument is that the groping, the halting progress with a stick, also has its privileges. The blind person sees what the person with vision does not, because she moves tentatively. Because instead of making use of direct lines of vision to distant objects, she gropes her way across the terrain. But Appelbaum argues that in the groping there is a kind of poise, what he calls a 'poised perception'. This is:

> a gathering unto a moment of novelty. It is perception of traces of hidden meaning. It is the perception that belongs to the stop.
>
> (Appelbaum 1995, 64)

Understood in this way, blindness implies a range of sensitivities and sensibilities to that which passes the sighted person by. Blindness is no longer a loss. Or if it is a loss, it is also a gain. I take my lesson here from Appelbaum. This is a book about method – and reality – that is also about the stop. The stop slows us up. It takes longer to do things. It takes longer to understand, to make sense of things. It dissolves the idea, the hope, the belief, that we can see to the horizon, that we can see long distances. It erodes the idea that by taking in the distance at a glance we can get an overview of a single reality. So the stop has its costs. We will learn less about certain kinds of things. But we will learn a lot more about a far wider range of realities. And we will, or so I also argue, participate in the *making* of those realities.

Method? What we're dealing with here is not, of course, just method. It is not just a set of techniques. It is not just a philosophy of method, a methodology. It is not even simply about the kinds of realities that we want to recognise or the kinds of worlds we might hope to make. It is also, and most fundamentally, about a way of being. It is about what kinds of social science we want to practise. And then, and as a part of this, it is about the kinds of people that we want to be, and about how we should live (Addelson 1994). Method goes with work, and ways of working, and ways of being. I would like us to work as happily, creatively and generously as possible in social science. And to reflect on what it is to work well.

Appelbaum writes that 'the danger of method is that it gives over to mechanical replacement' (Appelbaum 1995, 89). 'Mechanical replacement' has nothing to do with machines. Rather it has to do with the automatic. My hope is that we can learn to live in a way that is less dependent on the automatic. To live more in and through slow method, or vulnerable method, or quiet method. Multiple method. Modest method. Uncertain method. Diverse method. Such are the senses of method that I hope to see grow in and beyond social science.

The pleasures of reading

Why do the books fall into two heaps, the novels on the one hand, and the academic volumes on the other? Why do the novels get themselves read at the weekends, or on holidays, or in the ten minutes before falling asleep at night? Why do the work-books get read in the day, at prime times?

Then again, another kind of question. *How* do these different kinds of books get read? Why is it that reading a novel brings pleasure not only for its plot and its characterisation, but also for its use of words? If we reflect on the sheer pleasure of reading a well-crafted novel, one in which the words are carefully chosen, put together just right, then we may ask the question: what is the pleasure in reading an academic book? And how many academic books are really well written at the word-level? At the level of crafting?

How these two kinds of books get read is often, perhaps mostly, very different. If we read novels we read them, often, as an act in itself, for the pleasure of the read, the 'good read' of the airport novel, or the crafted text of a Barbara Kingsolver or a Penelope Lively or a J.M. Coetzee. They are pleasures in themselves, intrinsic. Whereas I guess we do not often read an academic book for the pleasure of the read itself, the pleasure, so to speak, of the journey. Rather we read it for the destination, where it will take us, where we will be delivered. We take pleasure, to be sure, in a well-crafted academic book – the ones that come to mind for me are, perhaps, mostly by historians. But the interest is different.

Perhaps, then, the distinction is between means and ends. Novels are ends in themselves, worth reading in their own right. Academic writings are means to other ends. The textures along the way, the actual writing, these are subordinate to those ends. It may be more agreeable to travel first class than third, but in the end we all arrive at the same destination.

What difference would it make if we were instead to apply the criteria that we usually apply to novels (or even more to poetry) to academic writing? Wouldn't the library shelves empty as the ranks of books disqualified themselves? What would we be left with? And, more importantly, if we had to write our academic pieces as if they were poems, as if every word counted, how would we write differently? How much would we write at all?

Of course we would need to imagine representation in a different way. Poetry and novels wrestle with the materials of language to *make* things, things that are said to be imaginary. It is the making, the process or the effect of making, that is important. The textures along the way cannot be dissociated from whatever is being made, word by word, whereas academic volumes hasten to describe, to refer to, a reality that lies outside them. They are referential, ostensive. They tell us how it is out there.

How, then, might we imagine an academic way of writing that concerns itself with the quality of its own writing? With the *creativity* of writing? What would this do to the referent, the out-thereness?

STS

Arguments from the discipline of STS (science, technology and society) will play an important part in the argument of this book. Thus, though I weave together a number of sources, the shifting ground on which I stand comes first and foremost from STS. A few words, then, on STS, and its role in the argument.

STS is the study of science and technology in a social context. The basic intuition is simple: it is that scientific knowledge and technologies do not evolve in a vacuum. Rather they participate in the social world, being shaped *by* it, and simultaneously *shaping* it. Some of the implications of this intuition are uncontentious. Who is going to deny the social significance of genomics or informatics, or try to argue that these are not shaped by social and economic concerns? Other implications are less obvious and much more controversial. Is the structure of current scientific reasoning patriarchal? Is the content of scientific knowledge at the same time essentially social? Does scientific theory and practice necessarily carry and enact social and political agendas? Is the distinction between scientific inquiry and knowledge on the one hand, and other forms of inquiry or experience on the other, a social contingency? Is the knowledge produced by scientists more or less contexted and local rather than possibly universal? Does scientific practice help to produce the objects that it describes and explains? Many STS scholars would answer 'yes' to each of these questions. But as is obvious, all, in greater or lesser degree, run counter both to common-sense and to many versions of the received philosophical wisdom.

So as I deploy the STS arguments (together with related positions developed in such disciplines as anthropology, sociology and cultural studies) these are going to take us to some more or less unfamiliar and sometimes anxiety-provoking territory. But this is precisely my object. STS work over a period of thirty years has made a series of strong and counter-intuitive claims about the character of science. These have profound potential implications for the conduct of natural science. But if they have implications for natural science, then so, too, they are potentially important for social science. So it is a source of some frustration that those arguments – and their implications – have not been more important in social science and its thinking about method and methodology. And such, to be sure, is the object of this book. I work through some of the STS findings in the context of social science, and in doing so attempt to destabilise some standard versions of social science wisdom. All this means that my argument moves between natural and social science. There are certainly important distinctions between the two, but here, for the most part, I trade upon their commonalities. These, I take it, are instructive and important. And this is where I start.

The argument outlined

In Chapter 2 I offer an account of a laboratory ethnography described by STS writers Bruno Latour and Steve Woolgar. The issue is: how is scientific knowledge produced? Their answer is: in a more or less messy set of practical contingencies. But what is most startling is their additional claim that in its practice science *produces* its realities as well as describing them. This is the cornerstone of my own argument. It runs counter to common-sense, and is also easily misunderstood, since it sounds as if it is a way of saying that 'anything goes'[13] and one can believe what one wants. But this isn't right. If realities may be built, Latour and Woolgar also show that it is difficult to do this. In practice bright ideas are very far from realities. And it is the word 'practice' that is the key. If new realities 'out-there' and new knowledge of those realities 'in-here' are to be created, then practices that can cope with a hinterland of pre-existing social and material realities also have to be built up and sustained. I call the enactment of this hinterland and its bundle of ramifying relations a 'method assemblage'.

But do those practices narrow down, converge, to make a single reality? In Chapter 3, I follow an account by Annemarie Mol of the practices of medical diagnosis, and argue that they don't. She shows that different practices tend to produce not only different *perspectives*, but also different *realities* – even for what otherwise might seem to be single-disease conditions. She calls this 'the problem of multiplicity'. But if there are different realities, then lots of new questions arise. How do they relate? How do we choose between them? How should we choose between them? One possibility is that we need what Mol calls an *ontological politics*. If truth by itself is not a gold standard, then perhaps there may be additional *political* reasons for preferring and enacting one kind of reality rather than another. Such, at any rate, is a possibility.

If realities made in methods are multiple, then do they have to be definite and fixed in form? Chapter 4 answers this question by saying 'no'. Using two more case studies – the treatment of alcoholic liver disease in the UK NHS, and a water pump in Zimbabwe – it shows how realities may change their shape or become more or less indefinite. But is this okay? Are the bush-pump or alcoholic liver disease not just definite objects that we haven't quite understood? Is our vagueness a sign of methodological failure? The answer is, perhaps, but I don't think so. Instead I argue that (social) science should also be trying to make and know realities that are vague and indefinite *because much of the world is enacted in that way*. In which case it is in need of a broader understanding of its methods. These, I suggest, may be understood as methods assemblages, that is as enactments of relations that make some things (representations, objects, apprehensions) present 'in-here', whilst making others absent 'out-there'. The 'out-there' comes in two forms: as manifest absence (for instance as what is represented); or, and more problematically, as a hinterland of indefinite, necessary, but hidden Otherness.

But if this is so, then how might we know about the indefinite or the non-coherent? Clearly we cannot know the indefinite without limit. It ramifies on for ever. But at least we can explore the issue, and this is the topic of Chapter 5 where I consider the character of allegory as a method for non-coherent representation. Again I work through cases. I argue that a rundown set of premises can be understood as an allegory for health-service disorganisation because it is tolerant of realities that are multiple, diffuse and non-coherent. Again, following work by Vicky Singleton, I suggest that the UK cervical smear programme is held together as much by inconsistency as consistency – that is by the ubiquitous practice of the allegorical. Finally, I argue that the horrors of a train collision can also be understood as a performative allegory for railway disorganisation – but also of pain and suffering. All of these are modes of knowing, methods assemblages, that do not produce or demand neat, definite, and well-tailored accounts. And they don't do this precisely because the realities they stand for are excessive and in flux, not themselves neat, definite, and simply organised. But this does not mean that they are not good methods.

So method assemblage works in and 'knows' multiplicity, indefiniteness, and flux. But how might we think about this? What *are* methods – or methods assemblages? In Chapter 6 I explore this issue by discussing materials from three very different sites of inquiry: management techniques, sociological ethnography, and religious experience. I argue that all of these are method assemblages because they detect, resonate with, and amplify particular patterns of relations in the excessive and overwhelming fluxes of the real. This, then, is a definition of method assemblage: it is a combination of reality detector and reality amplifier.

Chapter 7 returns to the question of truth and asks what follows if this is no longer a methodological gold standard. If it is no longer the only 'good'. Politics, we have seen, is another, 'good', but there are further possibilities. Others might include the aesthetic (beauty), and the spiritual or the inspirational. I develop this argument by looking at forms of method assemblage where there is little attempt to distinguish between such goods. Using materials drawn from Australian Aboriginal practices and the writing of Helen Verran and David Turnbull, I show that few Euro-American assumptions about representation and reality hold in Aboriginal cosmology. There is no universal reality. Realities are not secure but instead they have to be practised. And the world is not passive, waiting to be seen by people. Aboriginal cosmology both puts together goods that are usually held apart in Euro-American metaphysics, and it is explicit that all is enactment. To say this is not to say that science and social science practice should follow the Aboriginal model – but it shows once more that the metaphysics of method are, in principle, endlessly variable.

The argument of the book raises a series of more or less radical questions about method, and I review these in Chapter 8. I press for a more generous, and inclusive approach to method, and as a part of this briefly touch on a series of destabilising questions about the character and role of academic inquiry, and about knowledge more generally. This is because the division of labour which founds the academy, between the good of truth and such other goods as politics, aesthetics, justice, romance, the spiritual, inspirational and the personal, is in the process of becoming unravelled. This implies that we need to look not only at our practices but also at our institutions if we are to create methods that are quieter and more generous. Perhaps the model that we need, or one of the models, is that of 'partial connection' (Strathern 1991). At any rate, if the argument works at all then we need to find ways of living in uncertainty. The guarantees, the gold standards, proposed for and by methods, will no longer suffice. We need to find ways of elaborating quiet methods, slow methods, or modest methods. In particular, we need to discover ways of making methods without accompanying imperialisms.

INTERLUDE:
Notes on empiricism and autonomy

Euro-American common sense tends to the reflex that it is important to stipulate the conditions under which science can be properly carried out. This is because scientific inquiry needs to be protected from the distortions that might come from outside. The idea that science needs to be protected in this way is often (though not always) linked to 'empiricism' and to 'positivism'. *Empiricism* is a family of traditions in the philosophy of science which argue that scientific truths grow out of, and are properly generalised from, appropriate empirical observations. *Positivism* is another, closely related, set of traditions which argue that scientific truths are rigorous sets of logical relations or laws that describe the relations between (rigorous) empirical descriptions.

In the social sciences, empiricism and especially positivism are now usually seen negatively. Raymond Williams comments that positivism is a 'swear word by which nobody is swearing' (1988, 239). No doubt this is right. However their basic intuitions are widespread in Euro-American common-sense thinking about science and social science. It is commonly assumed that observations should be unbiased and representative, and that theories should be logical and consistent both with one another, and with observation.

The sociology of science, which was invented by Robert K. Merton (1973a; 1973b) started out on this assumption. There were good reasons for Merton's intuitions. He was writing at the time of Nazi racial science, and Stalinist Lysenkoism (which argued that plants could transmit and inherit acquired characteristics). He argued that these lethal lapses from proper scientific standards were a consequence of the failure to insulate science from political agendas in totalitarian societies. Scientists' capacities for unbiased observation and logical thinking were being eroded by these agendas. Instead science should, he said, be protected by a 'scientific ethos'. First, it should be *universalist*, testing its ideas in terms of: '*preestablished impersonal criteria*: consonant with observation and with previously confirmed knowledge' (Merton 1973a, 270). This meant that the race, gender, politics, or national origins of the scientist were not relevant to truth. Second, it should be *disinterested*. Scientific claims should be assessed independently of local social, economic, political, and personal interests. Third, it should be *sceptical*. Scientists should not take things on trust. (Merton talked of *organised scepticism*.) And finally, it should be *communal*. By this Merton meant that scientists should always publish their results: that science would best advance if it published its findings.

Merton's vision of science throws up some problems. (It is, for instance, difficult to see how scientists are consistently sceptical: in practice if they are to be effective they have to take a lot on trust.) And there are problems, too, with empiricism and positivism (we will encounter some of these below). But this is a convenient place to start because Merton is very clear that anything that interferes with 'empirically confirmed and logically consistent statements of regularities' (1973a, 270) is illegitimate because it detracts from the proper empirical and logical basis of truth. Merton's theory, then, is that research needs to be *disentangled* from the social and the psychological, and entangled solely with logic, with facts, and with methods for determining the facts.[14]

This is a language to which we will return. Different visions of science propose that it should be (or it is) entangled and disentangled with the world in different ways. Empiricism offers one recipe for this. It tells us that science (and social science too) have to be autonomous if they are to work properly. They should be disentangled from the social.

2 Scientific practices

... tools only exist in relation to the interminglings they make possible or
that make them possible. The stirrup entails a new man–horse symbiosis that
at the same time entails new weapons and new instruments. Tools are insep-
arable from symbioses or amalgamations defining a Nature–Society machinic
assemblage. They presuppose a social machine that selects them and takes them
into its 'phylum': a society is defined by its amalgamations, not by its tools.
Similarly, the semiotic or collective aspect of an assemblage relates not to a
productivity of language but to regimes of signs, to a machine of expression
whose variables determine the usage of language elements. These elements do
not stand on their own any more than tools do.

(Deleuze and Guattari 1988, 90)

A proposition, contrary to a statement, includes the world in a certain state.
... Thus a construction is not a representation from the mind or from the
society about a thing, an object, a matter of fact, but the engagement of a
certain type of world in a certain kind of collective.

(Latour 1997, xiii–xiv)

Inscription devices and realities

In October 1975 a young French philosopher arrived at the Salk Institute in
San Diego. Called Bruno Latour, he later wrote that his 'knowledge of science
was non-existent; his mastery of English was very poor' (Latour and Woolgar
1986, 273). He watched the work of the Salk Institute endocrinologists for
nearly two years and then wrote a book about it with sociologist of science
Steve Woolgar. Called *Laboratory Life*, this appeared in 1979 and, with books
by one or two others,[15] helped to create a new field, that of the *ethnography of
science*.

As we move through the present book we will look over the shoulders of
ethnographers as they visit scientific laboratories, clinics, hospitals, religious
ceremonies and managerial meetings. We will also watch the work of social
scientists – and others – as they produce knowledge in practice. So what do
ethnography of knowledge practices tell us? The answer is that ethnography
lets us see the relative messiness of practice. It looks behind the official accounts

of method (which are often clean and reassuring) to try to understand the often ragged ways in which knowledge is produced in research. Importantly, it doesn't necessarily distinguish very cleanly between science, medicine, social science, or any other versions of inquiry. Distinctions such as these tend to go out of focus in the welter of knowledge practices uncovered by ethnography. It also tends to find continuities between natural and social science. Physicists may have their instruments, but so too do sociologists. Much that we learn about the practice of natural science is also applicable to social science.

Thus the first take-home message from Latour and Woolgar is that what the authors called 'the tribe of scientists' (1986, 17) is not very different from any other tribe. Scientists have a culture. They have beliefs. They have practices. They work, they gossip, and they worry about the future. And, somehow or other, out of their work, their practices and their beliefs, they produce knowledge, scientific knowledge, accounts of reality. So how do they do this? How do they make knowledge?

The ethnographers of science are usually more or less *constructivist*. That is, they argue that scientific knowledge is constructed in scientific practices. This, it should be noted, is *not* at all the same thing as saying it is constructed by scientists. Thus we will see that practices include, and imply, instruments, architectures, texts – indeed a whole range of participants that extend far beyond people. But the process of building scientific knowledge is also an active matter. It takes work and effort. The argument is that it is wrong to imagine that nature somehow impresses its reality directly on those who study it if they just set aside their own biases. The picture of science offered by Merton is not right. But how is this construction done?

Different ethnographers respond to this question in somewhat different ways. However Latour and Woolgar, whom I follow here, explore it materially. They wouldn't call themselves 'materialists' because they do not think that everything derives from, or can be ultimately explained in, material terms. Nevertheless, they are very much into *materiality*. This means that they focus in the first instance on the physical stuff of the laboratory, and how this is laid out architecturally. For instance, it has a chemistry section, a physiology section, and then there is a location with desks and word processors which is mainly to do with paperwork. Then they talk about the way materials move around. Energy, money, chemicals, people, animals, instruments, tools, supplies, and papers of all kinds, move into the laboratory. At the same time, people and (different) papers and maybe instruments, together with debris and waste, move out. Looked at as a system of material production, then, the major product of the laboratory turns out to be *texts*. These are very expensive: at 1979 prices they cost about $30,000 each. No doubt the figure would be much higher now.

If the Salk Laboratory is a system of material production then how are its various material resources turned into texts? Latour and Woolgar trace this through a number of moves. Step one: they observe that 'the desk . . . appears to be the hub of our productive unit' (1986, 48). At the desk two kinds of texts are juxtaposed: on the one hand some come from outside the laboratory, such as scientific articles or books; on the other hand some originate from within the laboratory. But where do these come from? The answer is that they are produced by what they call *inscription devices*.

So this is the second step in their argument. An inscription device is a system (often including, though not reducible to, a machine) for producing inscriptions, or traces, out of materials that take other forms:

> an inscription device is any item of apparatus or particular configuration of such items which can transform a material substance into a figure or a diagram which is directly usable by one of the members of the office space.
>
> (1986, 51)

For instance, an inscription device might start out with rats. These would be sacrificed to produce extracts which would be placed in small test tubes. Then those test tubes would be placed in a machine, for instance a radiation detector, which would convert them into an array of figures or inscriptions on a sheet of paper. These inscriptions would be said – or assumed – to have a direct relation to 'the original substance'.

At this point, stage three, something interesting happens. Latour and Woolgar argue that *the process of producing the traces melts into the background*:

> The final diagram or curve thus provides the focus of discussion about properties of the substance. The intervening material activity and all aspects of what is often a prolonged and costly process are bracketed off in discussions about what the figure means.
>
> (1986, 51)

The argument is thus that *the materiality of the process gets deleted*. (Perhaps this is why 'constructivism' is often mistakenly thought to be about a purely human activity.) For what is subsequently manipulated is not the rats themselves. It is not even the extracts from the rats. Rather it is curves derived from figures from the relevant inscription devices. It is the curves that get juxtaposed with one another on the desks of the researchers.

The fourth step in the story is a process of isolating, detecting, and naming substances:

> Samples of brain extract are subjected to a series of *discriminations*. . . . This involves the use of some stationary material (such as a gel or a piece of blotting paper) as a selective sift which delays the gradual movement of the sample of brain extract. . . . As a result of this process, samples are transformed into a large number of fractions, each of which can be scrutinised for physical properties of interest. The results are recorded in the form of several peaks on graph paper. Each of these peaks represents a discriminated fraction, one of which may correspond to [a] . . . discrete

chemical entity. . . . In order to discover whether the entity is present, the fractions are taken back to the physiology section of the laboratory and again take part in an assay. By superimposing the result of this last assay with the result of the previous purification, it is possible to see an overlap between one peak and another. If the overlap can be repeated, the chemical fraction is referred to as a 'substance' and is given a name.

(1986, 60)

This is very important. Latour and Woolgar are telling us that it is *more or less stable similarities between curves* that allow the scientists to say that they have isolated a 'substance'. It is the relative similarities of successive curves that allow the laboratory workers to name a 'substance'. By contrast, 'elusive and transitory' substances – witnessed by curves that appear and disappear – come to be known as 'artefacts' and are disregarded.

Though some of their language is unusual, and, yes, they have taken us away from empiricism, perhaps what Latour and Woolgar have told us so far is not too surprising. But with the next step we move towards the unexpected:

The central importance of this material arrangement [of laboratory inscription devices] is that none of the phenomena 'about which' participants talk could exist without it. Without a bioassay, for example, a substance could not be said to exist. The bioassay is not merely a means of obtaining some independently given entity; the bioassay constitutes the construction of the substance.

(1986, 64)

'Without a bioassay, for example, a substance could not be said to exist.' And this is not simply a way of speaking. Here they are again:

It is not simply that phenomena *depend on* certain material instrumentation; rather, the phenomena *are thoroughly constituted by* the material setting of the laboratory. The artificial reality, which participants describe in terms of an objective entity, has in fact been constructed by the use of inscription devices.

(1986, 64)

This, then, is their fifth point. It is that *particular realities are constructed by particular inscription devices* and practices. Let me emphasise that: *realities* are being *constructed*. Not by people. But in the practices made possible by networks of elements that make up the inscription device – and the networks of elements within which that inscription device resides. The realities, they are saying, simply don't exist without their matching inscription devices. And, implicitly at least, they are also saying that such inscription devices – and even more so their particular products – are elaborate and networked arrangements that are more or less uncertain, more or less able to hold together, and more or less precarious.

As is obvious, this is an account of scientific inquiry that departs from the most common-sense – and indeed philosophical – understandings of the nature of reality and the ways in which we know it. It is certainly not empiricist: Merton, along with many others, assumes that there is a reality out-there of a definite form waiting to be discovered, if only we can get it right. But one does not have to be, an empiricist to feel that this is a good intuition. The same hunch underpins much more elaborate understandings of science – for instance the various versions of realism. So what does it mean to assert the contrary – to say that particular realities are *constructed* in networks of practices that include inscription devices and their contexts? What does it mean to say that without a bioassay a substance *could not be said to exist?* These are the puzzles that Latour and Woolgar leave us with. And they are puzzles central to the argument of this book.

A perspective on reality

Linear perspective. The art historians[16] tell us that this, known in antiquity and lost in the Dark Ages, was rediscovered in the early years of the fifteenth century by the Florentine architect Filippo Brunelleschi. In effect Brunelleschi asked himself the following question: is it possible to make a drawing of a building which looks exactly like the building itself? His answer, in an ingenious experiment with a mirror, was yes, it was.[17] With appropriate care a depiction could indeed reproduce the proportions of the object that it represented. The system of linear perspective so derived was developed and formalised by a further Renaissance architect, Leon Battista Alberti in his *Della Pittura* which appeared in 1435. The art, or the science, he told his readers, is to think of a picture as if it were a window, looking out in the direction of whatever is to be drawn. Or better, to think of it as an initially transparent screen, through which the external world can be seen.

But how to do this? Alberti makes two moves. The first is to imagine that there are lines of sight, coming from outside the window/screen, and passing through the screen to the eye of the painter. If the painter can mark the point where they pass through the screen on the way to his eye, then he or she will successfully mimic whatever is outside. The lumpy three-dimensional reality beyond the figurative window is thus converted into a two-dimensional representation. The first move, then, imagines a cone of vision starting, or ending, at the eye. Lines of sight beginning or ending in the eye, fan out, through the figurative window to the objects in the world beyond that window.

The second is to invent something that is usually called the vanishing point. The issue here is, how best to preserve the proportions of objects that are out there, in the world, when they are being transformed into a representation on a two-dimensional surface. Alberti suggests, in part, that we imagine a second cone, another fan. But this time, instead of converging in the eye of the artist, it converges on the *other side* of the picture/window, in the middle of the field of view, at a distant point on the horizon, directly opposite the eye of the artist. This, then, becomes the point at which those edges of objects in the real world that are at right angles to the picture/window tend to converge. To help in painting the artist now needs to draw this second cone, to depict it on the two-dimensional surface of the picture/window.

What form does this take? The answer is that it becomes a set of lines radiating out from a single point on the surface. This becomes the vanishing point. And the location of the vanishing point is fixed because it is where the line joining the centre of the two cones that have been created – the one converging on the distant vanishing point out there in the world, and the other, in here, in the eye of the artist – passes through the surface of the picture/window.[18]

This theorising is only a small part of the story. The conventions of linear perspective were being developed in the last years of the fourteenth century among artists in Italy. Art historians such as Norman Bryson (1983) show that it indeed took several generations for the new techniques to become established in the repertoire of the Renaissance artists.[19] This is partly because there were other powerful representational traditions available, for instance to do with the all-seeing eye of God, and symbolisms attached to various depicted features of nature or the gesture. Nevertheless it led to such powerful representations as Raphael's *Marriage of the Virgin*.[20]

Five assumptions about reality

To make sense of the stories about the Salk Laboratory and Western perspectivalism I need to talk about 'reality'. I need to talk about what is or isn't out there in the real world. That is, I need to engage with what philosophers variously call 'metaphysics' or 'ontology'. *Ontology* is the part of philosophy concerned with what there *is* and what there could be.[21] Philosophers talk of *metaphysics* when they are thinking about the untestable and often implicit assumptions that frame experience. From a philosophical point of view we all work in terms of more or less unexamined metaphysical (and ontological) assumptions. This is not a problem: there is no choice! But my interest is in the assumptions that these two stories make about reality, and in particular with Latour and Woolgar's surprising conclusion that specific realities are constructed in sets of practices that include particular inscription devices. At the same time, I am also interested in why it is that we might find this thought surprising.

In order to think about this I want to tease out some of the metaphysical assumptions that Euro-American people tend to carry when they, when we, think about what it is that scientists or social scientists are up to in the world. Or lay people. When we think, in other words, about reality, about what is, about ontology.

First, and most generally, it appears that our experience is widely if not universally built around the sense that there is, indeed, a *reality that is out there* beyond ourselves. Note that if we assume this then we are not committing ourselves to anything very specific. Indeed, I have phrased this in a way that is deliberately both general and diffuse. The out-thereness could take a variety of different forms. Let's think of this as a 'primitive' or 'originary' version of reality and simply talk of it as *'out-thereness'*. But for most Euro-Americans, at least most of the time, the sense of reality we carry is considerably more specific. So what does this include? Here are some additional suggestions:

Most of us would, I guess, implicitly commit ourselves to the further sense that this external reality is *usually independent of our actions and especially of our perceptions*.[22] Note that this – I will call it a commitment to *'independence'* – is not the same thing as out-thereness in its primitive form. As I have just noted, at least in principle, out-thereness might be experienced as much more closely related to our perceptions and our actions, much more dependent on them, than is generally the case in Euro-America.[23] I say 'generally' because there are at least parts of contemporary science – quantum mechanics is an example – in which the reality in question is taken to be closely related to any attempt to measure it.

Another more or less related common-sense is that this external reality comes before us, that it *precedes us*. Again this is not the same as the primitive commitment to out-thereness. It is a possible version or specification of it – but alternatives can be imagined. One could imagine, for instance, a theology or a metaphysics in which out-thereness was only possible in relation to a knowing and sentient being, or perhaps a set of methods for detecting and apprehending that reality. Versions of this, which are usually taken to be philosophically idealist (though this may be only one of the possibilities), have been considered from time to time in Western metaphysics. But, aside perhaps from some physicists in their professional lives, Euro-America mostly doesn't sense things that way. I will call this particular version of out-thereness *'anteriority'*.

A further common-sense is that external reality has, or is composed of, a set of *definite forms or relations*. Again, this is not entailed in the primitive commitment to out-thereness. Rather it is a possible operationalisation or version of it. One might, for instance, live in a world in which what went on was always vague, diffuse, uncertain, fluid, elusive and/or undecided – and was taken to be so. But though the social world may sometimes be apprehended in this way, Euro-American empirical experience mostly doesn't work like this. Instead it buys into an assumption that the world is more or less specific, clear, certain, definable and decided. It agrees, to be sure, that we may dream or imagine in ways that are vague and indefinite – but this has little to do with reality. It also agrees that individuals or groups may be vague and unclear (or simply mistaken) about the character of that world: our methods for finding out about it may be underdeveloped, distorted or themselves be vague. But this is usually seen as a failure in the attempts of those involved to gather proper knowledge, rather than being an attribute of the world itself. This I will call the assumption of *'definiteness'*.

Another common-sense is that the world is shared, common, the *same everywhere*. Once again, this is not implied in the primitive commitment to 'out-thereness'. Different people, groups or cultures might exist in different worlds. One could imagine multiple versions of the real (which is not the same thing as multiple perspectives on the same reality). Indeed this possibility is sometimes entertained, perhaps in a somewhat metaphorical form, in the context of social life, with the idea that different people live in different 'social worlds'. But nonetheless, again some parts of physics excepted, this would not be a common Euro-American intuition with respect to the physical world, or indeed in the end in the social world. Instead most Euro-Americans would be committed to what I will call *'singularity'*.

It is easy to think of other possible forms or specifications of the real. It is tempting, for instance, to think of *constancy* as a further category. (Do objects or processes in general stay the same unless they are disturbed? Most Euro-Americans would probably say yes.) Another, to which I will return at the end of the book, is *passivity* (in Euro-American versions of the real, the latter is usually 'disenchanted' and rendered passive). Yet another, though perhaps it does much of the same work, is *universalism*. But this initial list will do for the moment, because it allows us to distinguish between (a) Albertian perspectivalism, (b) Latour and Woolgar's understanding of scientific inquiry at the Salk Institute, (c) the scientists' own apparent understanding of their work (which is probably not so very far from that of Merton), and (d) our own possible surprise at the conclusion proposed by Latour and Woolgar.

First, then, Albertian perspectivalism. To work within this is surely to be committed to the entire list. *Out-thereness* of course: perspectivalism precisely depends on a distinction between observer and observed. *Independence?* Though perspectivalism has also been an imaginative and creative tool for Western artists for at least five hundred years, in the first instance (think of Brunelleschi) the issue is to find ways of representing the world out there. *Anteriority?* Again, thinking of Brunelleschi, it has to do with the representation of a pre-existing world. It assumes that there is a world out there already in place that is waiting to be depicted. *Definiteness?* Yes again. Importantly, the apparatus of per-spectivalism articulates a most specific and precise version of what it is to be definite. Thus the system is a projection that rests on the assumption that the real world is a Euclidean space, and that space is populated with representable objects possessed of Euclidean volumes. The art, or the science, is to discover and follow the rules that allow the relevant definite three-dimensional volumes to be transcribed on to a two-dimensional surface. And finally *singularity?* Again yes – and again linear perspective has its own particular take on this. If space is Euclidean, and it is populated with objects with specific volumes, then it follows that representational eyes in different places will see different views or perspectives. At the same time, since the rules are explicit, they precisely provide for the projection of a single three-dimensional real-world object from several different perspectival viewpoints. Perspectivalism is thus most strongly committed not only to a specific version of definiteness, but also, and as a part of this, to a specific and spatially-based version of singularity. Knowledge of the world resides in the subject.

So much for perspectivalism. Its version of out-thereness is highly specified. But what of the scientists in the Salk Laboratory? Look at this snippet of conversation between two of the Salk scientists as reported by Latour and Woolgar in *Laboratory Life*:

> *Dieter*: Is there any structural relation between MSH and Beta LPH?
> *Rose*: It's well-known that MSH has parts in common with Beta LPH. . . . Would you have expected finding proteolytic enzymes in the synaptosome?
> *Dieter*: Oh yes.
> *Rose*: Well, has it been known for a long time?
> *Dieter*: Well yes and no . . . there is a paper by Harrison showing that they do not obtain.
>
> (1986, 160)

Like any other conversation, this can be interpreted in various ways. However, the most straightforward reading suggests that Rose and Dieter, like the Albertian artist, are committed to and assume all five of the features of reality mentioned above. *Primitive out-thereness?* Yes. MSH and Beta LPH are only two of the external entities that appear in the conversation. *Independence?* Yes, each of these compounds is taken to have features independent of the beliefs, ideas, or practices of the scientific community. *Anteriority?* Yes, they pre-exist any attempt to get to know them. *Definiteness?* Yes indeed, that is what the conversation is all about. MSH, Beta LPH and proteolytic enzymes are all assumed to have definite attributes. The difficulty Rose and Dieter are wrestling with in the second part of the conversation doesn't call this into question: it is rather that the definite features of the enzymes appear to be in doubt amongst the relevant scientists. And finally, *singularity?* Again, yes of course. MSH is an object. It is a single object. It is a single object that can be compared with Beta LPH. It is not, it cannot be, different things in different places.

So Rose and Dieter are committed to a set of assumptions about reality very similar to those articulated in Euclidean perspectivalism. The only difference is in the way in which definiteness and singularity are detected. In perspectivalism they are specified in geometrically spatial terms, while endocrinological definiteness and singularity are generated in an alternative, chemically defined, manner.

But what of Latour and Woolgar? What of *their* assessment of the practices of the scientists? What of their counter-intuitive conclusion that particular realities do not exist without sets of practices that include inscription devices and the networks within which these are located? To tackle these questions we need to return to the Salk Laboratory.

The hinterland

Latour and Woolgar insist that science has to do with the *manipulation of inscriptions and statements*. As I have already noted, the desk of the Salk scientist, so central to scientific production, is covered with texts. Some derive from local inscription devices, and others from beyond the laboratory – papers, reviews and preprints written by scientists elsewhere. So the argument is that texts are put together and played off against one another. And the purpose of all this? It is to produce statements that carry authority, that tell about the outside world.

What do these statements look like? Latour and Woolgar divide them into a number of categories. Some are unconditional. They simply describe the outside world without qualification. For instance: 'Ribosomal proteins begin to bind pre-RNA soon after its transcription starts' (1986, 77). And, closely related to these, there are statements that are hardly statements at all because everyone takes them for granted anyway. These are only made explicit when talking to students or outsiders. Then, and usually (though not always) with less authority, there are statements that include what Latour and Woolgar call *modalities*. Modalities are qualifications or contexts that turn up within the text. They may be references to authors or to the way in which the statements were produced:

> [T]his method has *first* been described by Pietta and Marshall. If Pietta and Marshall have a strong reputation this might add to the strength of the claim.

However other modalities tend to undermine credibility:

> Recently Odell [ref.] has reported that hypothalamic tissues, when incubated . . . would increase the amount of TSH.
>
> (1986, 78)

The words *'has reported'* suggest an agnosticism about this claim which is therefore seen as uncertain. Yet other modalities turn statements into mere speculations or possibilities, and are even more erosive:

There is also this guy in Colorado. They claim that they have got a precursor for H . . . I just got the preprint of their paper.

(1986, 79)

A lot of the time, then, scientists are comparing statements of differing degrees of strength, selecting and playing them off against one another in the process of trying to create unqualified statements. The practice is similar to the comparisons between the curves produced by inscription devices. We have seen that if these map on to one another it may become possible to say that a 'substance' has been discovered. It may be possible to give it a name. It is the same with the relations between statements and their modalities. Similarities, overlaps, stabilities, repetitions, or positive relations between statements tend to increase their authority. If all goes well it may become possible to make statements that assert unqualified claims about substances and realities, pin these down, fix them, and make them definite. But this is only one possibility. In practice, Latour and Woolgar suggest that most statements are qualified and uncertain. Never achieving a modality-free existence, their speculative lives tend to be more or less brief.[24] Overall then, in the Salk Laboratory:

> The aim of the game was to create as many [unqualified] statements . . .
> as possible in the face of a variety of pressures to submerge assertions in
> modalities such that they became artefacts. . . . the objective was to
> persuade colleagues that they should drop all modalities used in relation
> to a particular assertion.

(1986, 81)

This form of words suggests that science is a literary exercise. It is about the fate of statements as they interact with one another. This is not exactly wrong, but it is also misleading because, crucially, *science is not just a literary exercise.* Natural (and social) science works with statements of a particular provenance. Thus statements do not idly freewheel in mid-air, or drop from heaven. They come from somewhere. Thus we can all dream up wish lists about the character of reality, but without support from other statements or inscriptions of an appropriate provenance they do not go very far. So we might put it this way: *if a statement is to last it needs to draw on – and perhaps contribute to – an appropriate hinterland.* But what is the nature of that hinterland?

We already have a partial answer for science. A part of the hinterland of a statement is other related statements. Is it consistent with these? Do they tend to support it? If the answer is 'yes' then they tend to add to its authority. But we have also seen that this is only a part of the story. Scientific statements also draw more or less directly from a network or a hinterland of appropriate *inscription devices.* Do the practices in which these are embedded produce figures that can be compared and tend to reinforce one another? If the answer is 'yes' then the authority of a statement increases. If it is 'no', then the statement is likely to enter the limbo of the might-have-beens. This, then, is the most important point: it is the character of this hinterland and its practices that determines what it is to do science, or to practise a specific branch of science. To a first approximation, then, science is an activity that involves the simultaneous orchestration of a wide range of appropriate literary *and* material arrangements. It is about the orchestration of suitable and sustainable hinterlands.

Inscription devices: Latour and Woolgar are canny in the way they use this term. An inscription device may be, but is not necessarily, a technology or an instrument. More generally, it is a set of arrangements for labelling, naming and counting. It is a set of arrangements for *converting relations from non-trace-like to trace-like form*. It is a set of practices for shifting material modalities. This is their understanding of the special materiality of science. It is the process of *making* particular kinds of relations in an experimental and instrumental set-up, and turning these into traces. This is why they insist that:

> We do not wish to say that facts do not exist nor that there is no such thing as reality. In this simple sense our position is not relativist. Our point is that 'out-there-ness' is the *consequence* of scientific work rather than its *cause*. We therefore wish to stress the importance of timing.
>
> (1986, 182)

The practices of science make relations, but as they make relations *they also make realities*. This is why Latour and Woolgar are interested in timing. Beforehand things are not clear and the realities in question are not yet made. Afterwards they are.[25] This means that scientific work is both robust *and* insecure. Its insecurity, typically invisible to outsiders, is apparent to anyone who visits a laboratory or knows anything about the actual conduct of science. As I have noted, things go endlessly wrong. This radiation counter is not calibrated, those rats are ill, or the new serum samples are odd. The deliveries of oxygen have been held up. And even (and perhaps more tellingly) when everything is going well experiments tend to produce traces that contradict one another and erode rather than strengthen putative accounts of reality. The future of reality is always at risk in a sea of uncertainty. It is extremely difficult to build stable relations in the laboratory. It is extremely difficult to build relations that will produce more or less stable traces.

Here is Latour describing himself stumbling round the laboratory:

> He had to remember in which beaker he had put the doses, and made a note, for example, that he had put dose 4 in beaker 12. But he found that he had forgotten to make a note of the time interval. With pipette half-lifted, he found himself wondering whether he had made a note before or after the actual action took place; obviously, he had not made a note of when he had made a note! He panicked and pushed the button of the Pasteur pipette into beaker 12. But maybe he had now put *twice*

the dose into the beaker. If so, the reading would be wrong. He crossed out the figure.

(1986, 245)

Methodical procedures and meticulous note-keeping are necessary. Otherwise a day's work is lost. (Lest it be thought that Latour was particularly clumsy let me add that I was responsible for similar minor debacles in the course of my own laboratory ethnographies.) So the practices of science are quite obsessively textual. Labelling, naming, writing down, noting – they are fixated on the business of keeping tabs on things. And if this fails then the work of the laboratory also fails.

The precariousness of the process of producing stable traces about stable realities is also witnessed by another well-documented feature of laboratory science: *the fact that it is often surprisingly difficult to reproduce the novel findings of one laboratory in other laboratories*. It is not uncommon that a statement generated from the inscription practices in one laboratory cannot be reproduced elsewhere.[26] Is this a cause for suspicion? Is the new claim about reality doubtful? The answer is yes to both questions. If statements do not map on to one another, if the patterns do not repeat themselves, then the realities they report are being undermined. It comes to look as if the statement reported not a fact but an artefact. But what does this mean? Answer: if the creation of facts is a relational activity – a question of assembling and fine-tuning the appropriate inscription devices – then it is equally possible that what is happening is a failure in such fine-tuning. If this is the case then it may be that there is need for more training, new and special equipment, the production of particular test samples (Salk Institute work was crucially dependent on these), the specialist manual skills of a particular experimentalist or technician, or the competence of an in-house computer programmer. If people can be trained or travel, if the precise experimental set-up can be reproduced, if novel equipment can be built – in short, if the relations in one laboratory can be configured in another – then the reality in question may be reproduced. As Latour and Woolgar bluntly put it:

In no instance did we observe the independent verification of a statement produced in the laboratory. Instead, we observed the extension of some laboratory practices to other arenas of social reality, such as hospitals and industry.

(1986, 182)

Or, even more pithily:

. . . if you carry out the same assay you will produce the same object.

(1986, 183)

If this is not possible, if 'the same assay' is not carried out, then the reality disappears into a limbo of questionable modalities.

This, then, is the implication of Latour and Woolgar's argument. Contrary to Euro-American common sense, they are telling us that it is not possible to separate out (a) the making of particular *realities*, (b) the making of particular *statements* about those realities, and (c) the creation of *instrumental, technical and human configurations and practices*, the inscription devices that produce these realities and statements. Instead, *all are produced together*. Scientific realities only come along with inscription devices. Without inscription devices, and the inscriptions and statements that these produce, there are no realities.

Where does this leave 'out-thereness'? We've seen that Latour and Woolgar treat this as the '*consequence* of scientific work rather than its *cause*'. But the implications of their argument are now clearer and we can return to the list of out-therenesses:

Independence: is external reality independent of our perceptions and actions? The answer is: it depends on what we mean by 'our perceptions and actions'. For individuals or particular sites of scientific production the answer is – largely – yes. It is difficult to imagine circumstances in which we could imagine, perceive, or act realities into being individually, or in our work. In that sense the outside world is independent of us. But collectively and in the longer run the answer is different. This is because particular realities are brought into being with and through the arrays of inscription devices and disciplinary practices of natural and social science. Reality, then, *is not independent of the apparatuses that produce reports of reality*.

Anteriority: does external reality precede our reports of it? The answer, again, is that it depends. In general the answer is no, it doesn't. Reality and the statements that correspond to it are produced together in the disciplinary and laboratory apparatuses of inscription. But in specific circumstances (and we are all, and all the time, in specific circumstances), there is always also a large hinterland of inscription devices and practices already in production. This means that an equally large hinterland of statements, and realities that relate to those statements, are already being made. There is a backdrop of realities that cannot be wished away.

Definiteness: does external reality come as a set of definite forms and relations? Again, the answer is both yes and no. Where statements fit together and reinforce one another the corresponding objects are named and acquire a definite form. Where this does not happen they do not. And, as Latour and Woolgar show, though the aim of the game is to make definite statements that correspond to definite realities, much of the time scientific inquiry deals with uncertainty, fuzziness and undecidability. An example: Latour and Woolgar describe the way that for a seven-year period starting in 1962 there was uncertainty about the existence and the character of a substance of particular interest to the Salk endocrinologists which came to be known as TRF. This changed in a way that was scientifically unsatisfactory because it was fuzzy, vague, and shifting. There were doubts about its very existence. It was only after 1966 that it became possible to talk of 'TRF' as a substance – and the chemical character of that substance was only turned into a firm statement in 1969. The moral of the story is that sometimes things are definite and sometimes they are not.

Singularity: is the world shared, is it common, is there a single reality? For Latour and Woolgar the answer is 'yes', but only *after* the controversies have been resolved and the statements reporting on nature have become fixed, definite and unambiguous. Before this happens not only is reality indefinite, but at least at times of scientific controversy it is also multiple. Multiplicity is the product or the effect of different sets of inscription devices and practices, for instance in different laboratories, producing different and conflicting statements about reality. Nevertheless, the end point – difficult but in their view none the less sometimes achieved in science – is a single reality and a single authorised set of inscription devices.

In sum, Latour and Woolgar take us some distance from everyday Euro-American expectations about out-thereness. Reality is neither independent nor anterior to its apparatus of production. Neither is it definite and singular until that apparatus of production is in place. Realities are made. They are *effects of the apparatuses of inscription*. At the same time, since there are such apparatuses already in place, we also live in and experience a real world filled with real and more or less stable objects.

A routinised hinterland: making and unmaking definite realities

So why is scientific reality relatively stable, at least a lot of the time? Latour and Woolgar suggest that we might think about this in terms of *cost*. The argument is that undermining the relations embedded in received statements is expensive:

> the set of statements considered too costly to modify constitute what is referred to as reality. Scientific activity is not 'about nature,' it is a fierce fight to *construct* reality. The *laboratory* is the workplace and the set of productive forces, which makes construction possible. Every time a statement stabilises, it is reintroduced into the laboratory (in the guise of a machine, inscription device, skill, routine, prejudice, deduction, programme, and so on), and it is used to increase the difference between statements. The cost of challenging the reified statement is impossibly high. Reality is secreted.
>
> (1986, 243)

'Reality is secreted.' Notice that this posits a kind of feedback loop. Statements stabilise, and then recycle themselves back into the laboratory. This means that once they are demodalised, *yesterday's modalities become tomorrow's hinterland*. And, as a part of this they tend to change in their material form:

The mass spectrometer is the reified part of a whole field of physics; it is an actual piece of furniture which incorporates the majority of an earlier body of scientific activity.

(1986, 242)

So why and how do they change their material form? A part of the answer is that it is easier to produce statements about realities – easier to produce realities – when these take standardised and transportable forms. Latour and Woolgar talk of reification, but perhaps the notion of *routinisation* better draws attention to what is most important. We saw above that the practice of fitting bits and pieces together to produce more or less stable traces is a precarious business. Much goes wrong in laboratory science. But if machines and skills and statements can be turned into packages,[27] then so long as everything works (this is always uncertain) there is no longer any need to individually assemble all the elements that make up the package, and deal with all the complexities. It is like buying a personal computer rather than understanding the electronics, and the physics embedded in the electronics and assembling one out of components. Thus in the above example the field of physics that is the hinterland of the mass spectrometer can be taken for granted. It does not have to be rebuilt or even understood by those who use the instrument. One sociology of science literature talks of 'standardised packages'. This is the point: in this way of thinking all the reality-describing and reality-making of natural (and social) science practices surfs on more or less provisional standardised packages that are, form part of, or support, inscription devices and practices. At the beginning of this chapter I cited Latour:

A proposition, contrary to a statement, includes the world in a certain state . . . . Thus a construction is not a representation from the mind or from the society about a thing, an object, a matter of fact, but the engagement of a certain type of world in a certain kind of collective.

(Latour 1997, xiii–xiv)

Latour, here twenty years on, is talking about Isabelle Stengers's philosophy of science[28] (and his talk of propositions rather than statements is a small but potentially misleading change in vocabulary). But the overall argument remains the same. It is not a matter of words representing things. Words and worlds go together. Propositions (as he is now calling them) include realities – include a collective. Include and grow from what I am calling the hinterland.

Certain additional consequences follow. The hinterland produces specific more or less routinised realities and statements about those realities. But this implies that countless other realities are being *un-made* at the same time – or were never made at all. To talk of 'choices' about which realities to make is too simple and voluntaristic. The hinterland of standardised packages at the very least shapes our 'choices'. We who 'choose' embody and carry a bundle of hinterlands. Nevertheless there are a whole lot of realities that are not, so to speak, real, that would indeed have been so if the apparatus of reality-production had been very slightly different.

A further and related implication is that the hinterland produces certain *classes* of realities and reality-statements – but not others. Some kinds of standardised inscription devices and practices are current. Some classes of reality are more or less easily producible. Others, however, are not – or were never cobbled together in the first place. So the hinterland also defines an overall geography – a topography of reality-possibilities. Some classes of possibilities are made thinkable and real. Some are made less thinkable and less real. And yet others are rendered completely unthinkable and completely unreal.

The economic metaphor suggests that it is easier and cheaper to create new inscription devices, new statements and new realities by building on to the routinised black boxes that are already available. It also suggests that as the process goes along it becomes more and more difficult and expensive to ignore or to undo the routines and create others and alternative realities. Latour and Woolgar again:

> Once a large number of arguments have become incorporated into a black box, the cost of raising alternatives to them becomes prohibitive. It is unlikely, for example, that anyone will contest the wiring of the computer . . . or the statistics on which the 't' test is based, or the name of the vessels in the pituitary.
>
> (1986, 242)

For individual practitioners it is often, perhaps usually, best to borrow from and make use of a very extensive and expensive set of inscription devices, because these would be extremely costly to overturn. Latour and Woolgar offer an example of this:

> when Burgus used mass spectrometry to make a point, he made it difficult to raise alternative possibilities because to do so would be to contest the whole of physics. Once a slide has been shown with all the lines of the spectrum corresponding to one atom of the amino acid sequence, no one is likely to stand up and object. The controversy is settled. But if a slide is presented which shows the spots of a thin-layer chromatography, ten chemists will stand up and assert that 'this is not a proof'. The difference, in the second case, is that any chemist can easily find fault in the method used.
>
> (1986, 242)

It is also a practical point for working scientists in another way too. Should they build on a particular standardised package or, alternatively, raise the stakes and the costs, go against the grain, and try to reorganise the hinterlands to generate one that is new? This is not a possibility open to most practitioners, even in the most straightforward economic terms. The money and the time to undo (say) the physics that lies behind mass spectroscopy and build an alternative set of inscription devices with their corresponding reality-statements and realities is not likely to be available.

In this argument, it is the hinterland of scientific routinisation, produced with immense difficulty and at immense cost, that secures the general continued stability of natural (and social) scientific reality. Elements within this hinterland, even sections of it, may be overturned (perhaps this is what Thomas Kuhn, whom we will touch on below, meant when he talked of 'scientific revolutions'). But overall and most of the time Latour and Woolgar are telling us that it is the expense of doing otherwise that allows the hinterlands of scientific reality to achieve relative stability. So it is that a scientific reality is produced that holds together more or less. That appears to be – and in a real sense is – independent of our particular scientific perceptions and actions. That appears to – and in a real sense does – predate those actions, is anterior to them. That is, indeed, definite. That is, in this account, singular – though the issue of singularity is one to which I will return in the next chapter.

A note on Foucault: limits to the conditions of possibility?[29]

The apparatuses of scientific (and arguably of social science) production produce something akin to what Michel Foucault described as the conditions of possibility. If we go with the economic metaphor then they set necessary limits – more or less permeable, but nevertheless limits – to those conditions.

So how does the present argument differ from that of Foucault? One answer has to do with empirical scope. Foucault and his interpreters insist that there is endless possibility for variation and creative innovation within the existing conditions of possibility.[30] Nevertheless, it is also well known that Foucault argued that the current conditions of possibility were established at the end of the eighteenth century in a set of strategies laid down within what he called the modern episteme. The argument is that at the beginning of the twenty-first century we are still being produced by that episteme and its conditions of possibility.

This may or may not be right. However, the picture of natural (and social) science production proposed by Latour and Woolgar and other STS scholars is drawn on a smaller scale. Perhaps there are larger limits set by modern disciplinary strategies that lie within and are being enacted by the different inscription devices and practices of modern natural and social science. But Latour and Woolgar's suggestion is more modest. It is that the limits to scientific knowledge and reality are set by *particular and specific sets of inscription devices*. The relations between these become an empirical matter.

Given the flexibility of the modern episteme, the position is not
necessarily inconsistent with that of Foucault. Further it shares with
him the commitment to the idea that it is not simply knowledge of
realities, but also realities themselves, that are generated in the practices
of production. My question, and one to which I will return in Chapter
3, has to do with singularity. Latour and Woolgar tend to assume that
inscription devices (and so their hinterlands) mesh together fairly well.
This seems to me uncertain.

Covering up the traces

But then there is the great question: why doesn't it look that way? Why is it
not obvious that inscription devices produce not only the statements about
reality but also the realities themselves? How come people don't see that
'phenomena *are thoroughly constituted by* the material setting of the laboratory'
(1986, 64)? Why is it that reality is taken to be independent, anterior, definite
and singular? How come scientists are said to 'discover' a reality that is anterior,
definite, and all the rest?

Latour and Woolgar have given us the elements that we need to answer these
questions. Thus we have seen that the object of scientific practice is to make
unqualified statements about reality. All the qualifying modalities need to be
deleted. We have also seen that it is important to routinise statements by
turning them into taken-for-granted assumptions, instruments, or skills. The
more the hinterland is standardised and (at least in certain respects) the more
it is concealed, the better.

But this means that as the modalities disappear, so too do almost all of the
processes in which statements and realities are produced. The largest part
of the work that has gone into their production is deleted. In the end, the
inscription devices themselves disappear, though those that are most novel are
likely to retain a foothold in the 'methods section' of scientific papers. But it
is the 'subjective' and the 'personal' that disappears first. The traces and the
statements in the laboratory are used 'in such a way that all the statements
were seen to relate to something outside of, or beyond, the reader's or author's
subjectivity' (1986, 84).

This deletion of subjectivity is crucial. In natural and social science research
statements about objects in the world are supposed to issue from the world
itself, examined in the proper way by means of proper methods, and not from
the person who happens to be conducting the experiment. If this is not
achieved, then independence and anteriority are not achieved either. If the
scientist appears in her text, if she appears as a person, then this undermines
any statement about reality.

So what is the consequence of this process of deletion? Latour and Woolgar
suggest that scientific statements should be seen as 'split entities':

On the one hand, it is a set of words which represents a statement about an object. On the other hand, it corresponds to an object in itself which takes on a life of its own. It is as if the original statement had projected a virtual image of itself which exists outside the statement.

(1986, 176)

So there is deleting and splitting. But then something else happens to complete the effect: there is a causal reversal or inversion. It is no longer the case that the manipulation of inscriptions is seen to produce particular realities. Instead it is the *realities that come first*:

Before long, more and more reality is attributed to the object and less and less to the statement *about* the object. Consequently an inversion takes place: the object becomes the reason why the statement was made in the first place.

(1986, 177)

This is the way in which reality becomes the determining factor. It is no longer the processes of comparing, contrasting, and weighing up inscriptions that produce reality. It is no longer the long sequence of actions, events and negotiations in which appropriate inscription devices are brought together and arrayed. Least of all is it the uses made of the special skills of particular technicians or programmers. It is not arguments, debates, discussions or controversies that *produce* reality. It is not the work that lies behind those debates and discussions. Rather it is *reality* that settles any disagreements. It is reality that produces statements.

The thing and the statement correspond for the simple reason that they come from the same source. Their separation is only the *final stage in the process of their construction*.

(1986, 183)

The result is a sense of a world that is out-there in far more than the primitive or originary sense. It is an out-thereness that is also assumed to be independent of and prior, anterior, to our scientific attempts to know it. It is assumed to be definite – even if we do not yet know that definite form because we have not acquired the methods we need to know it. And it is assumed to be singular.

Latour and Woolgar's proposal, then, is that this bundle of out-therenesses can be understood as an accomplishment rather than something that defines and sets limits to the ways in which we can properly know the world. Indeed, it is that out-thereness is better understood as an accomplishment rather than something given in the order of things. In short it is that the embedded hinterland of scientific method, the practices that it carries, work to *produce* a reality that is independent, anterior, definite and singular.

This is the bottom line of their ethnography of science. The hinterland of methods *enacts* realities. And (one can turn this round) those realities then enact the conditions of possibility of further research. They do not do so wantonly. They do not do so randomly. This is not a matter of the will, of lust, of desire, or of political visions. Nothing can be made real without the ramifications of an appropriate hinterland. But none the less, realities are enacted. If this is difficult then this is because it questions the self-evidence of Euro-American metaphysics; because it undermines the necessity of the methods that we happen to have available to us; because it presents us with possibilities (a reality *enacted?*) that are dangerous and potentially destabilising at least in principle not only to the metaphysics in which our methods are embedded, but also to the particular realities which they produce.

The method assemblage

Latour and Woolgar's proposal is that out-thereness is accomplished or achieved rather than having a prior and determinate form of its own. Realities are produced along with the statements that report them. The argument is that they are not necessarily independent, anterior, definite and singular. If they appear to be so (as they usually do), then this itself is an effect that has been produced in practice, a *consequence* of method. This suggestion flies in the face of most Euro-American metaphysics, including the more standard versions of the philosophy of science and social science.

Confronted with this claim we have a choice. We might opt to stick with a standard version of metaphysics. We could insist that the argument is wrong, and that whatever is out there is (at least usually) independent, anterior, definite and singular. If we take this line then it follows that we should continue to design our research methods along the current lines. We will need to think of our methods as tools for discovering a reality, or aspects of a reality, that is out there in a fairly definite form but is more or less hidden to us. This is comfortable, reassuring, and fits many understandings of methods. However, there are good reasons for considering the less conventional alternative: that the metaphysics are not right.

There is much that might be said about this. Here are a few thoughts. First, even though its argument is unfamiliar, it is plausible. Even if it doesn't fit the standard Euro-American justifications, Latour and Woolgar's account fits the *practices* of natural and social science. The findings of their ethnography are neither empirically weird nor theoretically strained. They explain perfectly well why scientists (and social scientists and lay people) tend to be committed to a strong version of out-thereness. But at the same time they also show how this is consistent with the idea that out-thereness is something enacted in practice. As I have shown above, scientists are caught up in a hinterland that has both been created and yet is relatively obdurate because it is too difficult to overturn.[31]

Latour and Woolgar's argument applies just as well to our social sciences.

We too have our instruments of research. We too reflect on and work within the obdurate realities produced by the hinterland of those instruments. For instance, statistics do not exist *sui generis*. As is obvious, they have to be created. Indeed there has been considerable historical work on the way in which this has been achieved over a couple of centuries or more through the medium of elaborate systems of tallying, measuring and quantifying in such forms as censuses, timekeeping (or time-making), surveying and economic data-creation. Such apparatuses, the hinterlands of much of social science, embed and enact many assumptions about the nature of the social. Arguably, 'the social' was brought into being in these apparatuses, as they developed and carried strategies of social and state control. By now however, with so many daily practices (public and private) dependent on official and other statistics, their reversibility is in doubt. It is possible to tinker with them – but overall, undoing them would be extremely expensive both literally and metaphorically. The result is both that we have come to live, and are made, in a social reality that is partly quantitative in quite specific ways, and that much of this hinterland is bundled into and constitutive of social science research.[32] We might add that parts of it have also been produced by social science.[33]

None of this is to say that these statistics are wrong. They may be criticised for this or that particular failing, but this is not the point. Rather they and the relations in which they are located are hinterlands and social realities out-there that both enable and constrain any work in social science. They set limits to the conditions of social science possibility. Overall, then, this is the first reason for taking the arguments of Latour and Woolgar seriously. Though their argument about enacted realities sounds counter-intuitive, it is consistent with our Euro-American intuitions that realities, natural and social, are pretty solid. To say that something has been 'constructed' along the way is not to deny that it is real.

Second, and just as important, their argument helps us to think differently and more creatively about method. In particular, the suggestion that specific forms of out-thereness are enacted and re-enacted makes it possible to think about which realities it might be best to bring into being. This, as I hope I have made clear, is not a simple or trivial question of choosing the version of out-thereness that happens to suit. 'Choice', if this is an appropriate term at all, is limited by the need to relate to and build appropriate hinterlands that will sustain statements about reality. Philosopher Isabelle Stengers puts the argument in slightly different terms:

> no scientific proposition describing scientific activity can, in any relevant sense, be called 'true' *if it has not attracted 'interest'*. To interest someone does not necessarily mean to gratify someone's desire for power, money or fame. Neither does it mean entering into preexisting interests. To interest someone in something means, first and above all, to act in such a way that this thing – apparatus, argument, or hypothesis . . . – can concern the person, intervene in his or her life, and eventually transform it. An

> interested scientist will ask the question: can I incorporate this 'thing' into my research?
>
> (Stengers 1997, 82–83)

So this is not a trivial matter. 'Interesting' is not necessarily easy. Nevertheless, the implications are profound. If out-thereness are constructed or enacted rather than sitting out there waiting to be discovered, then it follows that their truth or otherwise is only one of the criteria relevant to their creation. Politics in one form or another also becomes important. But the moment we acknowledge this we are faced with new questions. What kind of out-thereness are possible? Which are so embedded that they cannot be undone? Where might we try to undo or redo them? How might we try to nudge research programmes in one direction rather than another?[34] To bend a phrase, if we think in this way then reality is no longer destiny.

In the rest of the book I pursue this non-conventional option. The stakes for politics, but also for truth, are surely so high that it would be mistaken not to try to think these through. But if we are to do this there are at least two reasons why we need a better vocabulary for talking of method. The first has to do with *symmetry* and the second with the character of the *hinterland*.

As I indicated in the introduction, conventional talk of 'method' is closely associated with rules and norms for best practice. Indeed, though method is usually more than this, it sometimes becomes indistinguishable from lists of do's and don'ts. But if we want to think about more generous versions of method we need to think seriously about methods that ignore the rules. Here the sociologists of science are helpful. I will discuss their notion of 'symmetry' more fully at the end of Chapter 5. However, for the moment I just need to say that the idea of symmetry suggests that we shouldn't let our ideas about what is true or false (in science or anywhere else) affect how we look at our subjects. For instance, if we build our assumptions about the nature of good methods into our investigations of method then we are likely to come to conclusions that mirror those assumptions. We are likely to find that 'good methods' produce 'good results'. We will tend to reproduce the current workings of method. The alternative is to follow Latour and Woolgar. As we have seen, they disentangle the asymmetrical normativities of standard methods-talk ('this is good science, and this is bad') from their stories about how methods work in practice. In this respect their inquiry is symmetrical – but so too are the terms of their analysis. This, then, is the first reason for devising a new vocabulary.[35]

The second reason relates to the hinterland of method. I have argued that method and its out-thereness are made out of, and help to make, an appropriate hinterland. I have also suggested (and this is the important point) that the hinterland ramifies out for ever. This means that method extends far beyond the limits that we usually imagine for it. Going beyond laboratory benches, reagents and experimental animals, or questionnaires, interview design protocols, and statistical or qualitative data-analysis packages it extends into tacit knowledge, computer software, language skills, management capacities, transport and communication systems, salary scales, flows of finance, the priorities of funding bodies, and overtly political and economic agendas. The list is endless. All of these form a part of the hinterland of research. Its boundaries are porous and extend outwards in every direction. However, the problem is that the word 'method' doesn't really catch these ramifications. To take one instance, it doesn't catch the way in which discourses about 'users' have become integral to most funded research in the UK over the last twenty years; or the ways in which related assumptions about audit have been embedded in the practice of research. This, then, is the second reason why we need a new vocabulary. We need a way of talking that helps us to recognise and treat with the fluidities, leakages and entanglements that make up the hinterland of research. This would allow us to acknowledge and reflect not only on what happens in laboratories or in the offices of social scientists, but also in the missing seven-eighths of the iceberg of method.

In order to do this I propose a (partial) neologism. When I want to refer to method in this extended manner I will usually speak of *method assemblage*. I will return to and redefine this term several times in what follows, and especially in Chapters 3 and 5. However I will start by noting that the term 'assemblage' comes from the English translation of Deleuze and Guattari's *Mille Plateaux* (see the citation that begins this chapter).[36] Helen Verran and David Turnbull say that for these authors an assemblage:

> is like an episteme with technologies added but that connotes the ad hoc contingency of a collage in its capacity to embrace a wide variety of incompatible components. It also has the virtue of connoting active and evolving practices rather than a passive and static structure.
>
> (Watson-Verran and Turnbull 1995, 117)

Here Verran and Turnbull have caught exactly what is needed. An *assemblage* (without the method) is an episteme plus technologies. It is ad hoc, not necessarily very coherent, and it is also active.

In Deleuze and Guattari the English term 'assemblage' has been used to translate the French 'agencement'. Like 'assemblage', 'agencement' is an abstract noun. It is the action (or the result of the action) of the verb 'agencer'. In French 'agencer' has a wide range of meanings. A small French–English dictionary tells us that it is 'to arrange, to dispose, to fit up, to combine, to order'. A large French dictionary offers dozens of synonyms for 'agencement' which together reveal that the term has no single equivalent in English.[37] This means that while 'assemblage' is not exactly a mistranslation of 'agencement' much has got lost along the way.[38] In particular the notion has come to sound more definite, clear, fixed, planned and rationally centred than in French. It has also come to sound more like a state of affairs or an arrangement rather than an uncertain and unfolding process.[39] If 'assemblage' is to do the work that is needed then it needs to be understood as a tentative and hesitant unfolding, that is at most only very partially under any form of deliberate control. It needs to be understood as a verb as well as a noun. Here is Derrida (of course in translation):

> . . . the word sheaf seems to mark more appropriately that the assemblage to be proposed as the complex structure of a weaving, an interlacing which permits the different threads and different lines of meaning – or of force – to go off again in different directions, just as it is always ready to tie itself up with others.
>
> (Derrida 1982, 3)

Note that. A *'complex structure of a weaving'*. A *'sheaf'*. And here are Deleuze and Claire Parnet:

> In a multiplicity, what counts are not the terms or the elements, but what there is 'between', the between, a set of relations which are not separable from each other.
>
> (Deleuze and Parnet 1987, viii)

So assemblage is a process of bundling, of assembling, or better of recursive self-assembling in which the elements put together are not fixed in shape, do not belong to a larger pre-given list but are constructed at least in part as they are entangled together. This means that there can be no fixed formula or general rules for determining good and bad bundles, and that (what I will now call) 'method assemblage' grows out of but also *creates* its hinterlands which shift in shape as well as being largely tacit, unclear and impure.

But what is *method* assemblage? In Chapter 5 I will define this as the enactment or crafting of a bundle of ramifying relations that generates presence, manifest absence and Otherness, where it is the crafting of presence that distinguishes it as *method* assemblage. But I need to build towards this definition, so the work of Latour and Woolgar suggests a provisional and more specific possibility. Method assemblage may be seen as the crafting of a hinterland of ramifying relations that distinguishes between: (a) 'in-here' statements, data or depictions (which appear, for instance, in science and social science publications, and include descriptions of method); (b) the 'out-there' realities reflected in those in-here statements (natural phenomena, processes, methods, etc.); and (c) an endless ramification of processes and contexts 'out-there' that are both necessary to what is 'in-here' and invisible to it. These might range from things that everyone in question knows (how to do chromatography), through mundanities that no one notices until they stop happening (the supply of electricity), to matters or processes that are actively suppressed in order to produce the representations that are taken to report directly on realities (these would include the active character of authorship or the trail of continuities between statements and the realities that they describe).

INTERLUDE:
Notes on paradigms

Kuhn's book, *The Structure of Scientific Revolutions* is the best-known example of a body of work that drove a coach-and-horses through the empiricist and positivist vision of science.[40] Kuhn works by means of exemplary historical cases and his argument defies easy summary. For him exemplars are lessons on how to see and understand the world. To be a scientist is to work through cases. Quick accounts don't get to the heart of things at all. Talk and statements are only the tip of the iceberg. So if I say that three features of his account of science are important for us, though this isn't exactly wrong it is also at odds with the deeper sense of his story. Nonetheless:

First, scientists don't come to their work naïve but with a whole package which he calls a *paradigm*. This includes law-like generalisations, implicit assumptions, instrumental and embodied habits, working models, and a general and more or less implicit world-view. As I've just noted, it also includes exemplary applications of relevant models and theories. Scientific training, says Kuhn, is about learning to see chosen empirical circumstances in terms that fit how other paradigm-sharing scientists see them: to see particular circumstances as instances or applications of relevant models and theories. He writes that students

> regularly report that they have read through a chapter of their text, under-
> stood it perfectly, but nonetheless had difficulty in solving a number of the
> problems at the chapter's end. Ordinarily, also, those difficulties dissolve in
> the same way. The student discovers, with or without the assistance of his
> instructor, a way to see his problem as *like* a problem already encountered.
> (Kuhn 1970, 189)

You simply have to go and do the experiments and learn how to *see* them properly. Book-learning will not do.

Second, scientists are *puzzle solvers*. The world presents empirical and theoretical puzzles that can be solved by applying, adapting, and extending the paradigm. This, indeed, is what it is to be a scientist: a puzzle solver who is committed to this package, applies it, and extends it.

Third, very rarely paradigms fail. Systematic attempts to resolve some important puzzle do not work. If this happens for long enough then a sense of crisis develops that may lead to a '*scientific revolution*' in which one paradigm is replaced by another. But this is unusual. Most scientists are engaged in the creative and mundane process, puzzle solving.

Kuhn's account has many similarities with that of Latour and Woolgar (no surprise, for these authors come after and draw from Kuhn). We can see the Salk scientists as puzzle solvers who draw on and are entangled with a hinterland of more or less standardised instrumental, theoretical, and embodied resources. Furthermore, much of that hinterland is *tacit*: paradigms are embodied in craft skills, unspoken assumptions, and inscription devices.[41] Knowledge is not primarily an explicit set of statements and theories. It is a more or less inexplicit and indeterminate hinterland.

This means that the entanglements of Kuhn's picture of science are quite unlike those proposed by Merton. First, they are much less clear. Second, the empirical has a quite different significance because in Kuhn's way of thinking it *is not possible to make observations of nature in a neutral way*. Instead, what scientists observe, and how they observe it, is always tied up with their paradigm:[42] recognition of similarity in scientific observation is *acquired*:

> What is built into the neural process that transforms stimuli to sensations has the following characteristics: it has been transmitted through education; it has, by trial, been found more effective than its historical competitors in a group's current environment; and, finally, it is subject to change both through further education and through the discovery of misfits with the environment. Those are characteristics of knowledge, and they explain why I use the term. But it is a strange usage, for one other characteristic is missing. We have no direct access to what it is we know, no rules or generalizations with which to express this knowledge.
>
> (Kuhn 1970, 196)

Observations could not be neutral. They cannot be disentangled from the context of training or the process of puzzle solving which makes up the hinterland. So though *scientists solve real empirical puzzles* the reality they are dealing with is partially dependent on the paradigm itself. Out-thereness is not wanton or fickle. It cannot be created willy-nilly. But the particular forms it takes are more or less specific. Kuhn's vision of knowledge is pragmatic: paradigms are tools for handling out-thereness. But they also in part *enact* that out-thereness. So, though there are differences between Kuhn on the one hand, and Latour and Woolgar on the other, perhaps it is only pushing Kuhn's vision of science a little to say that specific versions of out-thereness are not independent and prior to the paradigm. And that if they are definite and singular they only become so in relation to a particular paradigm.

The structure of scientific entanglement is far removed from that of Merton.

3   Multiple worlds

Different sites

The picture of method starts to shift. The argument is no longer that methods *discover* and depict realities. Instead, it is that they participate in the *enactment* of those realities. It is also that method is not just a more or less complicated set of procedures or rules, but rather a bundled hinterland. This stretches through skills, instruments and statements (in-here enactments of previous methods) through the out-there realities so described, into a ramifying and indefinite set of relations, places and assumptions that disappear from view. So what follows from this? This is the issue that I tackle in the remaining chapters of this book. What are the realities that are made in method? What are the forms of the out-thereness? What realities, out-therenesses, *might* be made in method? How do in-herenesses get made, and what might they look like? How are different realities, different methods, and different in-herenesses entangled with one another?

The inquiry needs to be practical: an exploration of method-in-practice. So what happens to different methods in practice, and how do they relate to one another? To explore this question we move to a large university hospital, 'Hospital Z' in the Netherlands, and follow philosopher Annemarie Mol. Mol is watching the doctors and the patients as they work with lower-limb atherosclerosis. This condition is mundane, indeed prosaic, but it is also distressing, more or less painful and handicapping, and sometimes deadly. Mol's question is the following: what *is* it, lower-limb atherosclerosis? To answer this question she takes us to a number of locations, starting with a surgeon's *consulting room*:

> The surgeon walks to the door and calls in the next patient. They shake hands. . . . The patient, a women in her eighties, takes a chair at the other side of the desk, clutching her handbag on her lap. The doctor looks in the file in front of him and takes a letter out. 'So, Mrs Tilstra, here your general practitioner writes you've got problems with your leg. Do you?' 'Yes, yes, doctor. That's why I come here.' 'Tell me, then, what are those problems. When do you have them?' 'Well, what can I say? It's when I try

to do something doctor, move, walk, whatever. Like, I used to walk the dog for long stretches, but now I can't, I hardly can. It hurts too much.' 'Where does it hurt?' 'Here, doctor, mostly down here, in my calf it does. In my left leg.' 'So it hurts in your left calf when you walk. Now how many metres do you think you can walk before it starts hurting?' 'What can I say? I think it must be, well, some, not a lot, some 50 metres I guess.' 'Good. Or not good. Well. And then, can you walk again, then, after some rest?' 'Yeah, if I wait for a while, after that, yes, I can. Yes.'

(Mol 2002, 21–22)

Mrs Tilstra is describing a complaint which the medical professional call 'intermittent claudication'. This is intermittent pain on walking. She wouldn't have been talking to the surgeon unless she'd already talked to her general practitioner. And she wouldn't have talked to her general practitioner unless she had pain in her leg when she was out walking the dog. But – or so says Mol, following Latour and Woolgar – she didn't actually have a condition called 'intermittent claudication' until she presented herself at the surgery. Before that the pain was 'diffuse' (2002, 22). Mol continues:

This does not imply that the doctor brings Mrs Tilstra's disease into being. For when a surgeon is all alone in his office he may explain to the visiting ethnographer what a clinical diagnosis entails, but without a patient he isn't able to *make* a diagnosis. In order for 'intermittent claudication' to be practised, two people are required. A doctor and a patient.

(2002, 23)

Intermittent claudication calls for both a patient and a doctor. If it is to be enacted it needs to be crafted out of a story by the former and the embedded knowledge of the latter. Here we see the bundling of a hinterland. We also sense shades of Kuhn's scientists: the surgeon skilfully sees the similarity between Mrs Tilstra's case and all the other cases of intermittent claudication that he's seen before. Because sometimes the stories do not fit. For instance, Mr Zender also talks to the surgeon about pain in his legs, but this happens when he is sitting, not walking. Something is wrong with Mr Zender's legs, but not intermittent claudication. As the surgeon puts it: 'You may have pain in your legs alright. But there's nothing wrong with your leg arteries' (2002, 42).

The practice of intermittent claudication grows out of a specific hinterland that includes the story of the patient and the skill of the physician – and the latter includes a theory of its origin. This says that intermittent claudication is caused by inadequate supply of blood to the legs. This occurs when the legs need more oxygen, which is usually in exercise. And this, in turn, is caused by atherosclerosis, which is why Mr Zender's problems don't fit. But is that all? A story and a theory? Often not. Frequently the body of the patient is also important in the consulting room. It doesn't speak about intermittent claudication, but in the hands of the examining surgeon it may come up with corroborating evidence:

> . . . the vascular surgeon holds Mr Romer's two feet in his full hands to estimate and compare their temperature. He observes the skin. And with two fingers he feels the pulsations of the arteries in the groin, knee and foot.
>
> (2002, 25)

One foot is warmer than the other – a sign of a poor blood supply to the second. The skin is poor on the second too, a further sign. And the pulsations in the same leg are weak at the ankle, which is a third sign. So the body is important too. It is best if it corroborates the story of the patient, which it does for Mr Romer. But sometimes it doesn't. For instance, surgeons say that patients sometimes learn stories from the television or something that they heard at a party, and sound as if they have intermittent claudication. But then a physical examination produces warm legs with strongly pulsating foot arteries. This isn't common, but it can happen.

But let's stop at this point with the clinic and shift to a second site. A few floors down in the same hospital there is the *pathology laboratory*. This has a large fridge and, on the day Mol visits, the fridge contains a foot and the lower part of a leg. This was amputated the previous day and sent to the laboratory to assess the state of the blood vessels (2002, 33). So what does this mean in practice? The answer is, first, that the pathologist cut out pieces of the artery and put them into containers with preserving fluid. Then a technician decalcified the artery and sliced thin sections from it. Then she stained those sections, and fixed them onto glass slides (2002, 37–38). After this it was possible to examine them microscopically. Here is the pathologist talking with Mol as they look together through the microscope at one of the sections of artery:

> 'You see, there's a vessel, this here, it's not quite a circle, but almost. It's pink, that's from the colourant. And that purple, here, that's the calcification, in the media. . . . Look, all this, this messiness here, that's an artefact from that.' He shifted the pointer to the middle of the circle. 'That's the lumen. There's blood cells inside it, you see. . . . And here, around the lumen, this first layer of cells, that's the intima. It's thick. Oh wow, isn't it thick! It goes all the way from here, to there. Look. Now there's your atherosclerosis. That's it. A thickening of the intima. That's really what it is.' . . . And then he adds, after a little pause: 'Under a microscope.'
>
> (2002, 30)

The pathologist is talking about something like a more or less furred-up pipe. The scale in the pipe, the furring, is the thickened intima. In the textbooks and in the expertise of the physicians a 'furred' artery with a thickened intima impedes the flow of blood – in which case the story fits Mr Romer's examination described above. If the flow is impeded there is little or no pulse, and not enough oxygen is being carried into the afflicted foot. And it fits Mrs Tilstra's story too because intermittent claudication, pain on walking, is caused, as we have seen, by the lack of oxygen.

So lower-limb atherosclerosis is produced in the clinic and the pathology laboratory. But it is also enacted in the *radiology department*. Here the patient lies on a table, and a needle is inserted into the artery in the groin – a tense moment for the professionals, for things may go wrong. The needle is followed by a catheter. Then everyone apart from the patient retreats into a neighbouring room behind a lead screen, and two buttons are pressed. One injects X-ray opaque dye through the catheter into the artery and the other starts the X-ray machine which takes a series of pictures of the leg. If all goes well this produces a series of angiographic pictures (the technique is called *angiography*) which show a two-dimensional version of the lumen, the un-furred sections of the leg's vascular system. This is a visual representation of the places where the blood (and the opaque dye) can get. So the result is a bit like a route-map, that can then be displayed and discussed. Where, and how much, is the stenosis, the reduction in flow?

> Decision making meeting. The light box. A surgeon walks up to the angiography under discussion. 'How much did you make of this?' he asks the radiologists, his finger pointing towards a stenosis. '70%. Come on, that's not 70%. If you compare it with the earlier part there, if you take that bit as the normal part, up here, I'd say it's almost 90%, this lumen loss.'
>
> (2002, 73–74)

Like the pathology laboratory, the radiology department has its own methods and practices. Its hinterland includes: the X-ray machine; the dyes; the catheters; the lead screens; the surgical incision; the antisepsis; the sedated patient; the table on which he lies; and a whole lot more. But here the product is not a microscope slide. Instead it is an angiograph, another quite different version and visualisation of lower-limb atherosclerosis. It is another way of thinking about lumen loss, though this time it does not directly visualise the thickening of the intima. And it is a visualisation that can lead to the kind of debate cited above, for differences between estimates of lumen loss tend to be high.

Clinic, pathology laboratory, radiology department. Three locations. But here is a fourth which is another quite different way of detecting and locating the narrowing of blood vessels, the stenoses. Called duplex, this uses *ultrasound*. A small probe is pressed on to the skin of the patient above a blood vessel – though first it is necessary to find the blood vessel, and make sure that the probe is in good contact with the skin (a special gel is spread over the skin).

The probe emits ultrasound, and detects reflected ultrasound waves. The operator is looking for a Doppler effect, differences in the reflected wavelengths caused by blood flow. These appear as colours on a screen. In particular she is looking for variations in the speed of flow since (it is assumed that) blood will flow more quickly where the vessel is partially restricted (where the lumen loss is greatest) and more slowly where it is not. In practice she tries to compare velocities (usually peak systolic velocity, PSV) for a healthy and a partially occluded artery, in order to calculate a 'PSV-ratio' (2002, 55–57). Then she converts this, more or less controversially, into a figure for lumen loss:

> PSV-ratio smaller than 2.5: a stenosis smaller than 50%. PSV-ratio equal to or larger than 2.5: a stenosis larger than 50%. No sign: occlusion.
>
> (2002, 78)

This is another method with its own specific hinterland. But just as angiography differed from pathology, which in turn differed from the clinic, so duplex differs from angiography, having its own set of devices, skills, and people. The patient is prepared in a different way (and much less invasively than for angiography). Indeed, the physics built into the devices are different too, since electronics are supplemented by acoustics for duplex while a more or less nineteenth-century version of electromagnetic theory is built into angiography.

And then there is the *operating theatre*. Mol:

> It is a fat leg. Nurses have coloured the inside of the thigh yellow with iodine. The surgeon makes a sharp straight cut that opens up the skin. The fat underneath it is carefully separated by a resident. Blood repeatedly obscures the view. Tissues are used to absorb it. Small vessels are closed off with a small pin. Larger ones tied off with blue threads. Heparin is added to prevent the blood from clotting. . . . The entire cut is then widened with a . . . clamp. . . . Ah, finally, there is the artery. An orange plastic thread is put around to mark it. Then a similar search for the artery is repeated just above the knee. . . . The surgeon makes two incisions in the vessel wall. . . . With a knife the resident loosens the atheromatous plaque from the rest of the arterial wall where the artery is opened up. He then inserts the ring of a stripper around the plaque. . . . The stripper is moved upward. Slowly. When it finally arrives in the groin, the entire stretch of atheromatous plaque has been loosened. . . . With tweezers the surgeon draws it out. He drops it in a small bin. There goes the thickened intima. With lots of debris attached to it. Its bright white contrasts with the greyish artery.
>
> (2002, 90)

This is a description of endarterectomy, one of the surgical procedures for removing the thickened intima which causes arterial stenosis. It is, again, its own set of arrangements, its own method assemblage. The surgeons may use the angiography as a kind of route-map in order to decide where to make the incision. If this happens than the angiography and everything that produces it form one part of the surgical hinterland. But others include the skills of the surgeon, the various tools of his trade: scalpels, clamps, stripper (a remarkably crude instrument in the form of a ring attached to a stiff wire), heated pin, heparin, tissues, swabs, the apparatus of anaesthesia, and all the other elements of the operating theatre. So the bundled hinterland of the operating theatre turns the thickened intima into the form of white debris that can be dropped into a bin. Again, intima and the stenosis take their own particular form in the operating theatre. Forms both similar to and different from those in the other method assemblages.

A single story

So there are many sites: the clinic (which can be divided between the patients' complaints and the physical examination); the pathology laboratory; the radiology department; duplex; and the operating theatre. Mol describes at least five locations at which lower-limb atherosclerosis appears, and she could find more. She writes, for instance, of 'walking therapy', an alternative, non-invasive treatment for intermittent claudication.[43] She also visits the haematology department, where there is research into the formation of the plaque which leads to stenosis and lumen loss. But let us stop at this point. We have a large number of locations and each is its own method assemblage, its own set of health-related crafts and practices. And, if we follow the logic proposed by Latour and Woolgar, then we also need to add that *each of these method assemblages is producing its own version of atherosclerosis*: that there are *multiple atheroscleroses*. But what should we make of this startling conclusion? Are we happy to see the erosion of reality as singular?

If you ask the professionals, they usually *talk* about a single object, or about a set of objects and processes that fit together to produce a single reality. I touched on this above. Thus they say that long-term changes in the blood, perhaps partly due to diet and insufficient exercise, may lead to a slow build-up of atheromatous plaque and thickening of the intima. At first this has little effect, but at a certain point lumen loss becomes so great that blood flow is reduced. Then the patient experiences pain on walking – intermittent claudication – and he or she is likely to go to the doctor's surgery. When this happens other symptoms become visible – for instance the relative absence of pulsations, and the discovery of temperature differences between the legs. With further investigation additional symptoms become visible, and their locali-sation becomes possible. Duplex may be used to locate partial stenosis by measuring increases in the maximum speed of blood flow, or angiography to pinpoint partial or total stenosis.

So much for the story of origins and diagnosis. But what of treatment? In milder cases this may take the form of walking therapy.[44] Otherwise the alter-native is surgery. I have described one version of this above. Endarterectomy is an operation in which the offending plaque is physically stripped from the inside of the artery. But there are at least two further possibilities. In an operation called percutaneous transluminal angioplasty (PTA) the stenotic vessel is inflated from the inside using a device like a tiny inflatable balloon on the end of a tube which has been inserted into the vessel. The object in PTA is to push aside the plaque and increase the diameter of the lumen. A third possibility is to create a bypass round the sclerosis. And a fourth – necessary if the blood flow is so poor that there is risk (or the reality) of gangrene – is amputation. And it is only with this fourth form of intervention that the practices of the pathology laboratory become possible: cutting and preparing thin cross-sections of vessels with stenosis, and observing the growth of the intima and the loss of lumen under the microscope.

Differences in perspective

To write in this common, indeed habitual, way is a form of perspectivalism. It buys into, enacts, and presupposes a classic Euro-American version of out-thereness. An atherosclerotic reality out there is made *anterior* to, and *independent* of medical intervention. It is both *definite* in form and *singular*. With this framing of out-thereness the first task is to work out what reality *is*: for instance, the condition of Mrs Tilstra's leg vessels, and the location of stenoses in those vessels. Then, the second task is to intervene in a way that will help her. Surgeons and their medical colleagues are committed to a strong version of out-thereness: like perspectival artists, or the Salk Laboratory scientists, they assume that they are all addressing the same reality. And, to be sure, sometimes everything works out smoothly. Pain on walking, clinical examination, angiography, duplex, surgical intervention, and pathology – all may fit together to produce a single co-ordinated atherosclerosis.

Sometimes. But sometimes relevant practitioners instead find that they are faced with poorly co-ordinated realities. So what do they do then? Here is one example:

> The pathologist: 'You, since you're so interested in atherosclerosis, you should have been here last week. We had this patient, a woman in her seventies. She had renal problems. Severe ones too. So she was admitted. And the next day she died. Paff, from one moment to the next. The nephrologists were aghast, and so, of course, was her family. So we were asked to do an obduction. Her entire vascular system was atherosclerotic. One of her renal arteries was closed off, the other almost. It was a wonder her kidneys still did anything at all. It was hard to see where they got their blood from. And it was more or less the same for every other artery we took out: they were all calcified. Carotids, coronary arteries, iliac arteries: everything. Thick intimas, small lumens. And she'd never complained. Nothing. No chest pain, no claudication, nothing.'

(2002, 45–46)

Here is a second example:

> 'Here, look at this. Have you seen the pressure measurements of Mr Iljaz?
> It's unbelievable. I can't believe it. If you look at these numbers he can
> hardly have any blood in his feet at all. And he came to the outpatient
> clinic all alone, on his motorbike. Said he had some pain. I can't believe
> it. Some pain. On these figures alone I'd say here's someone who can't walk
> at all. Who's screaming.'
>
> (2002, 64)

And here is a third:

> [Mrs Takens's bypass] might be occluded, for the angiographic picture
> shows no dye beyond a critical point: the white stops abruptly. The duplex,
> however, still shows a peaking graph below this point. Flow. One of the
> radiology residents asks: 'In a case like this, when the angio says "closed"
> and the duplex says "open": what should one believe?' Two surgeons,
> speaking with a single voice, say: 'Duplex'.
>
> (2002, 83)

These are examples of contradictions: between the results of the pathology
laboratory and the life of the patient (first case); between the life of the patient
and measurements from the clinic (second case); and between angiography and
duplex (third case). But what is their significance?

First, contradictions are important in the day-to-day practice of medicine.
For though medical professionals usually work with a strong, perspectival
version of out-thereness, this is only a means to the more important end of
intervening and helping the patient. Their major preoccupation is in working
out *what to do*. In an ideal world all the indications add up and fit together:
they are different but compatible in-here perspectives on a single out-there
reality. But since the world is not perfect often those involved need to work
out how to act in the face of conflicting indications. Their world is quite unlike
that of the Salk scientists. The latter are concerned with fixing reality, with
truth, but in practice if not in theory the medical professionals often have to
work with multiple possible truths.[45]

Second, this points to the need for judgements, and for rules of thumb for
making judgements. These grow out of past experience, research, conversation
and reading. The craft of surgery. For instance, why did the two surgeons both
say 'duplex' in the same breath in the quotation immediately above? Mol's text
continues so:

> And then one of . . . [the surgeons] tells how he once studied seventeen
> cases like this: patients whose angiography showed an occlusion while their
> duplex showed flow. In all seventeen cases duplex proved to be in line with

the findings upon operation. 'It was only seventeen cases, so I couldn't publish it. But there were no exceptions.'

(2002, 83)

Here surgery is being used as the 'gold standard' to determine the nature of atherosclerosis, and duplex is being treated as the better guide to reality. But Mol also shows that in many (perhaps most) other contexts it is angiography that is used as the 'gold standard' to determine the accuracy of duplex (a much more recent technology) rather than the other way round. The implication is that that what counts as the best depends on circumstances (2002, 56). Rules, as Wittgenstein (1953) long ago showed, do not suggest their own proper application.

So there are more or less variable and situated rules for discriminating between contradictory versions of atherosclerotic reality, and deciding, in practice, what that reality is. For fixing it in practice. And, given the frequency of contradictions, such rules are endlessly deployed. How, for instance, could Mr Iljaz have come to the hospital outpatient clinic on his motorbike when he had so little blood in his feet? When he should have been screaming in pain? This is a puzzle:

'Yeah, that really is something' . . . nods [a senior internist], 'but we've seen cases like that before. Probably these people have only become worse very gradually. What happens is that their muscle metabolism alters. As long as people have time for it, the adaptation may go a very long way.'

(2002, 65)

This is one way of explaining away inconsistency, of turning it into *apparent* inconsistency. Another is the possibility that his diabetes may have led to degeneration of the peripheral nerves – in which case Mr Iljaz may not be able to feel pain in his legs. A third is that Mr Iljaz, who is an immigrant, may not speak Dutch well enough to explain how much pain he is actually feeling:

'Yeah, come to think of it, he may have underreported his complaints. His Dutch was poor.'

(2002, 66)

These are perspectives on difference that tend to *explain it away*. But what are the implications of this? It is tempting to say that the professionals are trying to cover up inconsistencies and even their incompetences. However this suggestion, common though it is in the literature of medical sociology, is surely only part of the story. More important in the present context is that such stories *help to sustain a strong perspectival and singular version of out-thereness* even as they manufacture multiple realities. They assume, and at the same time help to enact, the standard version of Euro-American metaphysics while also crafting something different. The implication is that in the present incomplete, uncertain, and untidy circumstances, we may not have full insight into either Mr Iljaz's particular condition or atherosclerosis in general. But nevertheless his condition, and the disease, are both visible out-there. They are out-there not just vaguely, but in all the different specifics we have discussed. They are *independent* of the investigations of medical science, they *precede* diagnosis, they are *definite*, and they are *singular*. It is a technical or practical matter if we are not yet properly clear of their attributes. In this way Euro-American metaphysics preserves itself in the face of possible contra-indications.

Multiplicity, enactment and objects

As I have hinted above, Mol wants to take us in another direction. Instead of singularity she is interested in difference and multiplicity:

> If a relation between the atherosclerosis of pathology and the atherosclerosis of the clinic is made, in practice, their objects may happen to coincide. But this is not a law of nature.
>
> (2002, 46)

Notice what is happening here. Mol is shifting the focus *from representation to the object itself*. Perhaps representations are being crafted too, but in-hereness is also a matter of objects, things. But why? And how? In what I have written above I have touched on the pivotal moment in her data. It happened at the moment in the pathology laboratory, when she was peering with the pathologist through the microscope at the cross-section of the artery from the amputated leg. Because what the pathologist said, I repeat the words, was: "'Look. Now there's your atherosclerosis. That's it. A thickening of the intima. That's really what it is. *Under a microscope*'" (2002, 30). Mol writes:

> My endeavour hangs on this last addition. The pathology resident utters it as if he is saying nothing special. 'Under a microscope'. But it implies a lot. Without this addition, atherosclerosis is all alone. It is visible *through* a microscope. A thickened intima. . . . There's something seductive about it: to use instruments as 'mere' instruments that unveil the hidden reality of atherosclerosis. . . . But when 'under a microscope' is added, the thickened intima no longer exists all by itself – but through the microscope. What is foregrounded through this addition, is that the visibility of intimas *depends on* microscopes. And, for that matter, a lot more.
>
> (2002, 30–31)

Objects, then, don't exist by themselves. They are being *crafted*, assembled as part of a hinterland. Like representations they are being enacted 'in-here', while sets of realities are being rendered visible out-there, and further relations, processes and contexts that are necessary to presence are also disappearing.

Unlike representations, however, objects do not *describe* the visible realities 'out-there'. This is method assemblage where the relations are different. Perhaps the in-here is being *made* by its visible out-there realities, or *caused* by them, or *shaped* or *given form* or *influenced* by them. (An atherosclerotic blood vessel might be caused by blood physiology, or influenced by poor diet, or both, depending on one's interest.) Then again, perhaps it is (also) having an effect, or shaping, or giving form or influencing the out-there. (Atherosclerosis enacted as angiography may have implications for the subsequent actions of surgeons or physical therapists.) For objects, then, the relations between the in-here and the visible out-there are complex, contingent and variable, and the traffic may be two-way.

If objects are enacted in this way, then this suggests that we need a second understanding of method assemblage to put alongside what has already been said about representation. We need to say that method assemblage may also craft hinterlands in the form of (a) in-here *objects*, (b) visible or relevant out-there *contexts*, as well as (c) out-there processes, contexts, and all the rest, that are both *necessary and necessarily disappear* from visibility or relevance. At the same time, however, if we focus on practice in this way then the perspectival pressure to singularity is weakened. And this is where the question of difference, of multiplicity, raises its head: when medicine talks of lower-limb atherosclerosis and tries to diagnose and treat it, in practice *at least half a dozen different method assemblages* are implicated. And the relations between these are uncertain, sometimes vague, difficult, and contradictory. This is what Mol calls the *problem of difference*. Because if we ruthlessly stick with the logic proposed by Latour and Woolgar, and pressed by Mol, then Euro-American perspectivalism will no longer do. We are not dealing with different and possibly flawed perspectives on the *same* object. Rather we are dealing with *different objects produced in different method assemblages*. Those objects overlap, yes. Indeed, that is what all the trouble is about: trying to make sure they overlap in productive ways. Ways that make it possible to intervene and help Mr Iljaz and Mrs Takens. So they overlap, *but they are not the same*. Different realities are being created and mutually adjusted so they can be related – with greater or lesser difficulty.

This is the point of Mol's intervention. Bar one subtle but devastating difference, her position is similar to that of Latour and Woolgar. And the difference? It is that medical inquiry and intervention *may* lead to a single reality, *but this does not necessarily happen*. In thinking of this Mol finds it helpful to distinguish between 'construction' and 'enactment':

> The term 'construction' was used to get across the view that objects have no fixed and given identities, but gradually come into being. During their unstable childhoods their identities tend to be highly contested, volatile, open to transformation. But once they have grown up objects are taken to be stabilized.
>
> (2002, 42)

Latour and Woolgar talk about *construction*. Their stories are full of talk about the vaguenesses of objects as they took (or failed to take) shape in the laboratory. They talk of the chosen few that made it through to the stable maturity of a perspectival 'closure'.[46] They add, as we have seen, that closure can in principle be undone, but also note that this is unusual because it is usually too expensive to undo the hinterland and remake it in some other form. TRF? The mass spectrometer? Closure has been achieved. The object has been constructed. A single hinterland is in place. No more questions.

So that is *construction*. But what of *enactment*? Mol:

> like (human) subjects, (natural) objects are framed as parts of events that occur and plays that are staged. If an object is real this is because it is part of a practice. It is a reality *enacted*.
>
> (2002, 44)

'Plays that are staged', writes Mol, pointing to the role of performance. But this is not an updated version of Goffman's dramaturgical sociology. Goffman distinguishes between *presentations* of self on the one hand, and *self* as a hidden reality lying behind and producing those presentations, on the other.[47] But this is precisely what Mol is trying to avoid. Her argument is much more closely related to recent writing in the philosophy, sociology and history of performance that emphasises the *performativity* of enactment than it is to Goffman's approach. It is these writings – in science studies, feminist theory, and cultural studies – which in one way or another have started to explore the possibility that there is a two-way traffic between enactments on the one hand, and realities on the other.[48] Enactments, it is being argued, don't just present something that has already been made, but also have powerful productive consequences. They (help to) make realities in-here and out-there.

To talk of enactment, then, is to attend to the continuing practice of crafting. Enactment and practice never stop, and realities depend upon their continued crafting – perhaps by people, but more often (as Latour and Woolgar imply) in a combination of people, techniques, texts, architectural arrangements, and natural phenomena (which are themselves being enacted and re-enacted). So Mol throws away the notions of construction and closure. Yes, of course, there are often *practical closures*. For the moment, in the present circumstances, and notwithstanding his apparent lack of pain, Mr Iljaz has really severe lower-limb atherosclerosis. But what there aren't are closures in general. Beware, Mol is telling us. If we attend to practice and to objects we may find that no objects are ever routinised into a reified solidity. We may find that there are no irrevocable objects bedded down in sedimented practices. We may find that the hinterlands are not set in stone. And if things seem solid, prior, independent, definite and single then perhaps this is because they are being enacted, and re-enacted, and re-enacted, in practices. Practices that continue. And practices that are also multiple. This is a way of thinking:

that does not simply grant objects a contested and accidental history (that they acquired a while ago, with the notion of, and the stories about their *construction*) but gives them a complex present, too, a present in which their identities are fragile and may differ between sites.

(2002, 43)

So Mol follows the lead suggested by Latour and Woolgar – but then shifts us in two ways. She moves from representations to objects, and as she does so, she also *does away with singularity*. In-hereness and out-thereness can be, and indeed usually are, *multiple*.

Virtual singularity

Here is a case of a single enactment that turns out to be multiple. It is the decision made by the British government in 1965 to cancel a warplane called the TSR2. This is an account of that decision by one of the participants:

> The discussion showed there had been a certain divergence amongst those concerned. James Callaghan, as Chancellor of the Exchequer, wanted to cancel the plane altogether for purely financial reasons. Ranged against him were (a) Denis Healey, who wanted to cancel the TSR2 and to substitute the American F111A, which would mean a certain saving of money but an enormous increase of outlay in dollars; and (b) Roy Jenkins, who wanted to cancel the TSR2 and replace it with a British plane – which was roughly George Brown's view as well; and (c) George Wigg, who held the view that we might have to cancel both but we mustn't make any decision until we had finished the strategic reappraisal which would show what kind of plane was required.
>
> (Crossman 1975, 190–191)

And here is a second account by another participant:

> But we had to have a decision, and the Cabinet was called again for 10.00 p.m. By midnight I had to resolve a difficult . . . decision. The Cabinet was split three ways: some favoured continuing with TSR2; some favoured its outright cancellation; and the third group supported the Defence Secretary's view that TSR2 should go but that its military role should be taken over by an order for American Phantoms, together with one for a number of F111As.
>
> (Wilson 1971, 89–90)

These accounts are not quite the same. Indeed, as the following table suggests, they only partially overlap:

	Crossman account	Wilson account
Cancel	✓	✓
Cancel and buy F111A	✓	
Cancel and buy British plane	✓	
Probably cancel but wait for strategic review	✓	
Continue with TSR2		✓
Cancel and buy F111A and Phantoms		✓

Euro-American common sense suggests that we should think about the differences perspectively. In this way the reality out-there is independent of, and prior to, the descriptions of that reality in-here. It is also definite and singular. If we think this way then we can assume that there was a single decision – and probably a single decision-making moment. At the same time, and as a part of the process of decision-making, we can assume that this was preceded by the elaboration of a number of specific options. But if we say this, then what should we make of the differences between the accounts?

The answer is that they are smoothed away. As with the differences between different versions of atherosclerosis, other explanations are found for the disparities. Perhaps those who wrote the accounts forgot what happened, or misunderstood it. Perhaps their accounts are self-serving. There are various possibilities, but they are all perspectival. They all preserve the assumption that there was indeed a single and definite decision made, selected from a single and definite range of options. Euro-American metaphysics in this way sustains itself.

But there is also the alternative multiple possibility, the proposal made by Mol. This is that the different participants were making *different* decisions, and that they simply thought they were making a single decision. Then, somehow or other, they co-ordinated themselves. Imagined themselves to be making the same decision. Displaced the possible differences, kept them apart. Perhaps we might call this 'virtual singularity'. But if we do so then the 'virtual' has nothing to do with cyberspace, but rather with the glass blocks in school optical experiments which seem to show pins located in places where they are not really to be found.[49]

Multiplicity and fractionality

If we are interested in multiplicity then we also need to *attend to the craftwork implied in practice*. Remember the reversal described by Latour and Woolgar. They said: one, that practices simultaneously produce statements about realities and the realities they describe; and two, that when the modalities disappear, the realities are suddenly turned into the causes of those statements. Perhaps this is right, but Mol is issuing a methodological warning. If we want to understand practice, and the objects generated in practice, then we need to make sure that we don't get caught up in that reversal ourselves. This is because it is misleading. Realities are not explained by practices and beliefs but are instead produced in them. They are produced, and have a life, in relations. So what we need is ethnography or what Mol calls praxiography:

> after the shift from an epistemological to a praxiographic appreciation of reality, telling about what atherosclerosis *is*, isn't quite what it used to be. For somewhere along the way the meaning of the word 'is' has changed. Dramatically. This is what the change implies: the new 'is' is one that is situated. It doesn't say what atherosclerosis is by nature, everywhere. It doesn't say what it is in and of itself, for nothing ever 'is' alone. To be is to be related. The new talk about what is, does not bracket the practicalities involved in enacting reality. It keeps them present.
>
> (2002, 53–54)

A praxiography allows us to investigate the uncertain and complex lives of objects in a world where there is no closure. Where, willy-nilly, there is no singularity. It allows us to explore the continued enactment of objects. And as a part of this, it allows us to investigate the multiplicity of those objects, the ways in which they interact with one another:

> Ontology in medical practice is bound to a specific site and situation. In a single medical building there *are* many different atheroscleroses. And yet the building isn't divided into wings with doors that never get opened. The different forms of knowledge aren't divided into paradigms that are closed off from one another. It is one of the great miracles of hospital life: there are different atheroscleroses in the hospital but despite the differences between them they are connected. Atherosclerosis enacted is more than one – but less than many. *The body multiple* is not fragmented. Even if it is multiple, it also hangs together. The question to be asked, then, is how is this achieved.
>
> (2002, 55)

And that is what the largest part of Mol's book is about: the practices that generate that apparently oxymoronic object, the body multiple. But what are they? Mol's work suggests that there are a number of ways in which differences are regulated. Some are more or less perspectival:

- Important is the idea that there is indeed only one body, so that any differences are a consequence of failures or limitations in practice. Mol calls this *'layering'*. Symptoms or diagnostic signs which may be at odds with one another are distinguished in negotiation from the underlying condition itself which is taken to be consistent. Crucial, then, to the body multiple is a continued faith in the body singular.

- A *single narrative* may also be important, a narrative that smoothly joins theories about the aetiology of atherosclerosis with its anatomical, physiological and diagnostic expressions. Expressions that are in turn linked to judgements about the possibility and desirability of particular interventions. The larger narrative, then, smoothes together a single coherent object that it describes and explains.

- *Translations* also help co-ordination of multiples. These are processes in which one thing is turned into another. We have come across a number of examples: angiographs were being (controversially) converted into percentages of lumen loss; and so too were PSV ratios.

- *Submission* is a hierarchical version of translation. We saw one context in which angiography (often the 'gold standard') lost out to duplex. So the lesson is that local hierarchies and submissions are important but these too are made and remade, and they are not necessarily consistent.[50]

- *Rationalisations* may have a crucial role too. They take the form of additional layers of narrative that explain apparent inconsistencies away – like Mr Iljaz's arrival at the hospital on his motorbike when he should have been screaming with pain.

These are ways of handling multiplicities that reconcile different atheroscleroses and patch them together into singularity. They are perspectival because, at the same time they also work to preserve a general commitment to ontological singularity. Mol points to further strategies that also sustain this general commitment but do so without producing a single atherosclerosis:

- *Mutual exclusion*: some things exclude one another. It isn't possible to take cross-sections of an artery from a leg vessel that is attached to a living patient. Conversely, legs that have been amputated cannot be cross-examined for complaints about intermittent claudication. Here the clinic and the pathology department exclude one another. 'The incompatibility is a practical matter. It is a matter of patients who speak as against body parts that are sectioned' (Mol 2002, 35–36). So many practices and the realities that they enact are parallels, alternatives, collaterals, streams of activity that never come together.

- *Creating different objects*: sometimes it is said that different practices are in fact producing different objects rather than conflicting versions of the same object. Mol describes a comparative study of PTA (stretching the stenosed vessel with the little balloon) and walking therapy. The result suggested that PTA improved pressure measurements but made no difference to the distance of onset of pain when walking. But for walking therapy it was the other way round. Does this mean that the findings are inconsistent? Possibly, but in practice those reporting the study said that the patient was suffering not from one, but two atheroscleroses: 'pressure-atherosclerosis' and 'walking-atherosclerosis'. Two singular objects replaced a single multiple.

- *Creating composite objects*: but if objects can be separated, they can also be recombined to produce composite entities. And this is what happened to these two different atheroscleroses. 'In the "criteria for success according to Rutherford" improvement is defined in a composite way. It is a combination of clinical symptoms and ankle–arm [blood pressure] index' (Mol 2002, 68, citing F. van der Heijden). This is one atherosclerosis with two parts. 'Addition', Mol observes, 'is a powerful way of creating singularity'.

- *Locating in different places*: finally Mol notes that 'Incompatibilities don't stop patients getting diagnosed and treated. Work may go on so long as the different parties do not seek to occupy the same spot. So long as they are separated between sites in some sort of *distribution*' (2002, 88). Separation may occur *over time* (patients move from one site to another and can't be in all of them at the same time (2002, 115)). It may occur between *different patients* (who are operated on in different and mutually incompatible surgical ways). It may occur as *mutual recognition* (the distribution between atherosclerosis as a gradual process of deterioration, and its reality as a serious condition in the here-and-now (2002, 116)). Or finally, separation may occur by acknowledging *differences in the conditions of possibility*. (Surgery is necessary at present, but future work in haematology will hopefully prevent the development of atherosclerosis and surgeons will find, as they did with stomach ulcers, that they are no longer needed.)

There are many ways of reconciling difference and avoiding multiplicity. Some are perspectival, and others are not. Together, however, they work to push the possibility of multiplicity off the agenda. Rendered invisible, it becomes a part of the out-there that is arguably necessary to the practices in question but cannot be acknowledged. By contrast, if we attend to practice we tend to discover multiplicity. But here is another important point. We discover multiplicity, *but not pluralism*. For the absence of singularity does not imply that we live in a world composed of an indefinite number of different and disconnected bodies, atheroscleroses, hospital departments, or political decisions. It does not imply that reality is fragmented. Instead it implies something much more complex. It implies that the different realities *overlap and interfere with one another*. Their relations, partially co-ordinated, are complex and messy:

[The term atherosclerosis] . . . is a co-ordinating mechanism operative in conjunction with the various distributions. It bridges the boundaries between the sites over which the disease is distributed. It thereby helps to prevent distribution from becoming the pluralizing of a disease into separate and unrelated objects.

(Mol 2002, 117)

I cited Mol similarly above: '*The body multiple* is not fragmented. Even if it is multiple, it also hangs together' (2002, 55). Hinterlands partially intersect with one another in complex ways, and the practices bundling those hinterlands together generate complex objects. We will, I think, need a range of different metaphors if we are to start thinking this well, but here is a first possibility. Perhaps we should imagine that we are in a world of *fractional objects*. A fractional object would be an object that was more than one and less than many. The metaphor draws on an elementary version of fractal mathematics. Thus a fractal line is one that occupies more than one dimension but less than two. Perhaps, indeed, when we visit the hospital (or anywhere else) we are in a world of fractionality. We are in a world where bodies, or organisations, or machines are more than one and less than many. In a *world* that is more than one and less than many. Somewhere in between.[51]

Partial connections

The difficulty of the talk of fractionality suggests that we are pressing up against the conditions of possibility. The dominant enactments of Euro-American metaphysics make it very difficult to avoid singularity on the one hand, and pluralism on the other. Either there is a single world, or there are lots of different worlds. This is what seems to be the choice.

If there is a single world, then in science or in social science it is our duty to try to provide an account of it. And if there are lots of different worlds? Then we are faced with two alternatives. Either the different worlds are components of a single larger world, in which case, once again, it is our duty to try to provide an account of it. Or, alternatively, the different worlds have nothing to do with one another. And if they have nothing to do with one another? Then we are confronted with what are usually taken to be the horrors of relativism. So what are those horrors?

These come in three closely related versions. First there is *epistemological* relativism. This says that the knowledge in your culture is just as good as the knowledge in my culture, so there are no grounds for claiming that my account of out-thereness is any better than yours. Second, there is *ethical* relativism. This says that ethics are situated and local, and there are no grounds for claiming that my ethical standards are any better than yours. And third, there is *political* relativism. The argument takes the same form again: there are no reasons for preferring my politics over yours. We should live and let live.

Notice, though, what is happening here. We are being pressed, all the time, to make a choice between singularity and pluralism. Either there is one, one reality, one ethics, one politics, or there are many. There is nothing in between. This pressure to dualist choice is why I take it that we are being pushed up against the enacted limits of Euro-American metaphysics – and, to be sure, being asked to re-enact it. But the dualism imposed by the choice does not follow. Something in between is a possibility.

One way to see this is to think in empirical mode and ask the question: how far do arguments carry in practice? How far, for instance, do arguments about claudication carry? Are they only valid in the place that makes them? In one cultural location? Or do they travel universally? Setting the choice up like this, as an empirical version of the epistemological dualism, reveals that the choice is forced. For the empirical and matter-of-fact answer is that arguments about claudication travel so far, but only so far. The same is the case for any other argument about out-thereness. How far does it carry? So far, but only so far. The overlaps, for instance, between the arguments made by Australian Aborigines and Euro-American technoscience are limited (we will discuss some of these in Chapter 7). The arguments carry only so far. But often enough they do carry in some measure. And an empirical version of the ethical question – or indeed the political question – leads us to similar conclusions. How far do our ethical or political arguments carry? Answer: they go some way, but only so far. It is indeed a commonplace that people disagree over what a good world would look like.

So how should we respond to this? There are three options. It is possible to insist on singularity, and insist that those who do not see it our way are suffering from impaired vision: that their empirical, ethical or political perspective on reality is flawed. To do so is to re-enact Euro-American singularity. Alternatively, it is possible to insist on pluralism, and the essential irreducibility of worlds, of knowledges, of ethical sensibilities, or of political preferences, to one another. This is the relativist response. But there is a third option, or a family of options, in-between. It is possible to observe, in one way very matter-of-factly, that the world, its knowledges, and the various senses of what is right and just, overlap and shade off into one another. That our arguments work, but only partially. That is how it is. But how to *think* this? How to think the in-between?

Here is a possibility. Feminist technoscience writer Donna Haraway, and following her, anthropologist Marilyn Strathern, talk of *partial connections* (Haraway 1991a; Strathern 1991). This is partly a matter of partial connections between different people – or different groups of people. But it is more complicated than this because it also has to do with partial connections *within the same person*. We do not, this is the argument, have single identities. Strathern notes, for instance, that Strathern-the-feminist is not the same as Strathern-the-anthropologist. They write in different ways in different circumstances and for different audiences. At the same time, however, neither are they entirely separate from one another. Strathern-the-feminist is *included* in Strathern-the-anthropologist. Strathern-the-anthropologist writes in a way that is informed by, but not reducible to, Strathern-the-feminist. And the same is the case the other way round.

Strathern's argument is informed both by her reading of (and exposure to) the indigenous cultures of Papua New Guinea, and by contemporary debates about identity politics: the realisation that political alliances which depend on single identities are usually counterproductive in Euro-America where most people are better understood as having multiple, shifting, and partially connected identities.[52] But the position is very close to one of the arguments made by Mol. The crucial word is *inclusion*. The argument is that 'this' (whatever 'this' may be) is included in 'that', but 'this' cannot be reduced to 'that'.

Another example. Mol shows that clinical diagnoses often depend on collective and statistically generated norms. What counts as a 'normal' haemoglobin level in blood is a function of measurements of a whole population. She is saying, then, that individual diagnoses *include* collective norms, though they cannot be reduced to these (Mol and Berg 1994). At the same time, however, the collective norms depend on a sample of clinical measurements which may be influenced by assumptions about the distribution of anaemia – though it is not, of course, reducible to any individual measurement. The lesson is that the individual is included in the collective, and the collective is included in the individual – but neither is reducible to the other.

It appears, then, that in practice there are plenty of partial connections, partial inclusions, partial relations. It also appears that these do not reduce to one another. Haraway:

> Irony is about contradictions that do not resolve into larger wholes, even dialectically, about the tension of holding incompatible things together because both or all are necessary and true.
>
> (Haraway 1991a, 149)

So there is inclusion, contradiction, and sometimes, if we follow Mol, co-operation too. But there is never collapse into singularity. And the arguments against identity politics are just as applicable to the out-therenesses of objects, of non-social realities.

Ontological politics

what is ceaselessly perfected is a history of erasure.

(Appelbaum 1995, 17)

By now we are well into this journey of erosion – the erosion of the self-evidences of Euro-American metaphysics and their versions of in-hereness and out-thereness. Realities, yes, they are real enough. Relativism is not an issue. One does not have to buy into Euro-American metaphysics to retain a commitment to out-thereness. So, yes, there is resistance. There is stuff. But the character of that stuff becomes less clear, less self-evident. Hinterlands are complex and ramified and only contingently coherent. Thus we have seen that out-thereness is not independent of practice in general, but only in particular. Though, of course, since we are all somewhere in particular, situated, we do not notice the distinction very much. We have seen that it is not prior to practice in general but only in particular. Though, again, since we are all somewhere in particular, we all live within a set of hinterlands of anteriority. Definiteness – I shall talk more carefully of definiteness in the chapters that follow, but again the formula applies. In general nothing is definite. Only in particular. And finally singularity. Mostly, yes, like the physicians and surgeons in Hospital Z, we find singularity. We *make* it. We live within it. But singularity likewise is very specific, very local. And it includes multiplicity.

So singularity is not only the product of specific enacted and visible out-therenesses – though their production of singularities is crucial – but also of a series of mechanisms for avoiding the appearance and the experience of multiplicity: for expelling it into invisibility. For, alongside the practices of multiplicity, there are endless practices for insisting on, presupposing, and producing singularity. There are stories about the singular nature of the world, its objects, and its processes. There are perspectivally inspired distinctions between (provisionally) hidden realities and appearances. There are processes for deleting the unfolding and uncertain nature of practices in favour of apparently stable and separate objects. There are methods for keeping different realities separate and distributing them across time and space. There are methods, we might say, for deferring multiplicity, for keeping it at arm's length, for effecting its disappearance. And, all the while, there is the practical business of reconciling multiplicities, of making the endless and complex ramifications of out-thereness look as if they were much more straightforward.

And such is the context and the character of Euro-American conditions of possibility. These talk of the necessity of singularity, but they also, and at the same time, *enact* multiplicity while erasing it, pushing it into invisibility. Atherosclerosis (or the decision to cancel an aircraft) are said to be single things but they are also being made multiply, or fractionally. So what should we make of this?

One argument is that the insistence on singularity is productive: that this *enables* invisibly multiple practices to craft invisibly multiple realities out-there. For instance, it may be that the idea that there is a single atherosclerosis makes it easier to create many different versions of the disease because it allows participants to assume that they are talking about (and making) a single condition. Or again, it may be that the idea that there is a single decision to be made about whether or not to cancel the TSR2 aircraft allows different participants to make multiple and different decisions. And then (for the argument can be looped back) it may be that it is the fact that multiple decisions are made which makes it possible to arrive at a single decision (for if everyone thought they were making different decisions, then it might be difficult or impossible to arrive at a single decision).

So there are arguments to be made for the current conditions of possibility for Euro-American out-thereness. Visible singularity, and invisible multiplicity. Perhaps this allows us the best of both worlds. Certainly Bruno Latour has made analogous arguments about the seeming (but only seeming) purity of modernity.[53] But there are counter-arguments too. In particular, it can be argued that presupposing singularity and deferring multiplicity into invisibility also makes it impossible to think about partial connections: to make visible the possibilities offered by what we might think of as *the discovery of fractionality*.

The possibilities? Yes. For the discovery of fractionality opens out the possibility that realities might be otherwise. This is like a deeper or broader version of the argument against the notion that biology is destiny. Notwith-standing its continued re-enactment in the popular and esoteric press, a large body of feminist-inspired writing demonstrates that biology is not destiny. And this is not simply because biological entities such as genes do not code for social behaviour. It is also, and much more profoundly, because biological entities are not themselves irrevocably fixed. Anatomical, endocrinological, genetic and expressive bodies are produced in different practices whose consistency – and indeed whose internal consistency – like atherosclerosis, is an uncertain product of moment-by-moment practice.[54] Biology is not fixed except in theory. It is enacted.[55] The problem, then, is that the commitment to visible singularity directs us away from the possibility that realities might in some measure *be made in other ways*. Or, to put it more generally, the presupposition of singularity not only hides the practice that enacts it, but also conceals the possibility that different constellations of practice and their hinterlands might make it possible to enact realities in different ways.

One way of putting this is to say that 'truth' is not and cannot be the only arbiter. In multiplicity or fractionality there are *varieties of truths*. But this then means that other values, concerns, and goods are also in play, one way or another, acknowledged or otherwise:

> The reality of atherosclerosis does not precede medical technology and the organization of health care, but is intertwined with them. This implies that the impairments of the body and the politics of crafting tools and organizing health care are intertwined as well. If this is so then reality, the physicalities or the psychology of a disease, cannot be the standard by which to assess treatments. The very advantages and disadvantages, the goods and bads, of *performing* reality in one way or another are themselves open for debate.
>
> (Mol 2000, 96–97)

Following in the footsteps of Foucault, Mol offers a provisional way of grasping at what we are after. Perhaps, in this alternative metaphysics of enacted fractionality we might think of what is made and what is told as an *ontological politics* (Mol 1999). Realities are real enough. These may take the form of in-here statements and the visible out-there realities they describe. This is what we learned from Latour and Woolgar. They may, as we have now learned from Mol, take the form of in-here objects or processes, and out-there contexts of one kind or another that go along visibly with those objects. Method assemblage, we are learning, needs to be about more than representations. But either way – whether we are talking about representations or objects, it becomes clear that truths are not the only arbiters. In an ontological politics we might hope, instead, to interfere, to make some realities realer, others less so. The good of making a difference will live alongside – and sometimes displace – that of enacting truth.

INTERLUDE:
Notes on interferences and cyborgs

Visions of science. Donna Haraway offers an important feminist version of method-and-politics that is also an ontological politics.

What, asks Haraway, would it be to be 'objective'? In standard practice the answer is usually detachment. Disentanglement from location. This is the kind of response offered by Merton, and by the empiricists and the positivists whom he follows. But, says Haraway, detachment is never possible. As we produce knowledges we are all located somewhere, in our practices and in our bodies. We are caught up, as she puts it, in a dense *material–semiotic network*. That is, we are caught up in sets of relations that simultaneously have to do with meanings and with materials.[56] We are entangled in our flesh, in our versions of vision, and in relations of power that pass through and are articulated by us. So detachment is impossible. At best a self-delusion, more often it is also a form of irresponsibility. It is irresponsible because it attempts what she calls the 'god-trick'. That is, it pretends to see 'everything from nowhere' (Haraway 1991b, 189). Whereas it is, indeed, somewhere. And it makes and remakes the textures of the material–semiotic networks.

How, then, to imagine objectivity? The answers vary (Daston 1999). But it seems that we cannot step outside, either to be neutral, or to find some special place – for instance a feminist or women's standpoint that sees further than the alternatives. But if we are always a part of what we explore then, writes Haraway:

> only partial perspective promises objective vision. This is an objective vision that initiates, rather than closes off, the problem of responsibility for the gen-erativity of all visual practices. Partial perspective can be held accountable for both its promising and its destructive monsters.
>
> (Haraway 1991b, 190)

But why *partial* perspective? We have touched on the answer above. Subjects, people, are not coherent.

> The topography of subjectivity is multi-dimensional; so, therefore, is vision. The knowing self is partial in all its guises, never finished, whole, simply there and original; it is always constructed and stitched together imperfectly, and therefore able to join with another, to see together without claiming to be another. Here is the promise of objectivity: a scientific knower seeks the subject position not of identity, but of objectivity; that is, partial connection.
>
> (Haraway 1991b, 193)

We are sets of partial connections. We are, to use the language that I am proposing, both in-here, as subjects, and out-there, as networks of meaningful and material relations. Or, to put it differently, people are, or form a part of, methods assemblages. In their envisioning practices they, we, bundle together not very coherent but nevertheless structured hinterlands. And objectivity, in the way Haraway redefines it, is possible if we acknowledge and take responsibility both for our necessary situatedness, and for the recognition that we are located in and produced by sets of partial connections.

Partial connection. This is the metaphor that lies behind Haraway's trope of the cyborg. Haraway:

> A cyborg is a cybernetic organism, a hybrid of machine and organism, a creature of social reality as well as a creature of fiction. Social reality is lived social relations, our most important political construction, a world-changing fiction.
>
> (Haraway 1991a, 149)

Cyborgs, then, are sets of partial connections. These may present themselves as political. They may present themselves as material (between machine and human, or between human and animal). And, as the citation suggests, they may present themselves as lying somewhere between reality and fiction. For, another visual metaphor, cyborgs are about interfering in the distributions between reality and fiction. Which is why Haraway picks up and plays with the metaphor of the cyborg. A product of the space age, a rarefied achievement of the military-industrial complex, Haraway seeks to tear the metaphor from its location of birth and bend it to interfere, to make interference patterns, in the unjust material–semiotic networks of what she calls 'the current disorder'. Indeed 'a world-changing fiction'. Reality and fiction relate to one another. They are included in one another. But they cannot and should not be reduced to one another.[57] So

> Feminist cyborg stories have the task of recoding communication and intelligence to subvert command and control.
>
> (Haraway 1991a, 175)

Multiplicity and partial connection. There is no gold standard. No single reality. Realities may be made and remade. They are made and remade. This is a version of ontological politics.

4 Fluid results

Sites

In this inquiry into in-hereness and out-thereness the argument of Chapter 2 was that method is productive of realities rather than merely reflecting them. And that parts of the out-there are made visible while other parts, though necessary, are pushed into invisibility. This was a first stage in the erosion of Euro-American metaphysical certainties. In Chapter 3, in a move that turns singularity into multiplicity, or better into fractionality, I explored the enactment of different realities at different sites. In Chapter 4 I extend the argument by considering a further metaphysical assumption, the idea that what is out-there necessarily has a definite form. I am going to say that there are circumstances where this is not the case: that some relevant realities are indefinite.

In this chapter I again work through case studies. The first is about the treatment of alcoholic liver disease in an area of the north of England. Vicky Singleton and I together collected the empirical materials and developed a version of the argument that follows, and I am grateful to her for her generosity in letting me rework that argument here.[58] The second is a study of a water pump in Zimbabwe by Marianne de Laet and Annemarie Mol.

The first case, alcoholic liver disease: I want to say that its diagnosis and treatment is similar to that of atherosclerosis because it too is located and enacted as an object in a wide range of different locations:[59] the Waterside District General Hospital, the alcohol advice centre, the Samaritans, the Salvation Army, the consulting rooms of the general practitioners, the residential homes – not to mention the pubs, off-licences and homes of those released from hospital back into the community. Alcoholic liver disease appears, and is enacted, in many places. However I'll start with the account of alcoholic liver disease in a *textbook* which sits in the office of one of the consultant gastro-enterologists in the Waterside District General Hospital:

> Over the last 20 years alcohol consumption has correlated with deaths from cirrhosis. . . . At the Royal Free Hospital, London, between 1959 and 1965 alcoholism accounted for only 4.3% of patients with cirrhosis, compared

with 25% between 1978 and 1983. . . . Not all those who abuse alcohol develop liver damage and the incidence of cirrhosis among alcoholics at autopsy is about 10–15%. . . . The explanation of the apparent predisposition of certain people to develop alcoholic cirrhosis is unknown.

(Sherlock 1989, 425)

The textbook tells us that the aetiology of alcoholic liver disease is complex, related (but not very directly) to: the level of alcohol intake; the duration of heavy drinking; patterns of drinking (bingeing is less dangerous than continued heavy drinking); gender (everything else being equal, women seem to be more at risk to alcoholic cirrhosis than men); histocompatibility; genetically determined enzyme polymorphism (there are at least two bodily metabolic systems for converting alcohol and how these balance off against one another can be important); exposure to hepatitis B; levels of dietary protein (malnutrition tends to exacerbate the effects of alcoholic liver disease); and to a variety of poorly understood metabolic synergies, for instance between nutrition and alcohol conversion (Sherlock 1989, 427–428). The mechanisms of liver damage are similarly complex, as are morphological changes in the liver, which come in at least three varieties: steatosis (fatty liver), alcoholic hepatitis (inflammation of the liver) and cirrhosis, which is 'a diffuse process with fibrosis and nodule formation. It has followed hepato-cellular necrosis' (Sherlock 1989, 410).

If we follow Latour, Woolgar and Mol, then we need to say that the textbook account of alcoholic liver disease both describes and helps to enact the condition in a particular way. Here the condition becomes a complex set of aetiological, environmental, physiological, anatomical and behavioural relations and effects which match the statements in the text. Indeed, the text uses the kind of apparatus described by Latour and Woolgar: paragraphs, figures, drawings, tables, and innumerable references. The text, then, is the in-here that belongs to and helps to enact a particular method assemblage. It crafts and is crafted in a hinterland that produces a somewhat tentative and uncertain description in-here of an alcoholic liver disease and its reality out-there.

Let's move to a second location. Again, we're in the office of a consultant gastro-enterologist, Dr Warrington, but this time we're interviewing him, and a somewhat different reality is being created:

> When they arrive my juniors are sat down, and I tell them how to manage liver disease. The instructions I give them are quite specific. They are told to follow a written protocol. Alcoholic liver disease can be quite easy to manage. But very few understand the basic principles.[60]

Dr Warrington tells us that he gets a 'little bit annoyed' if juniors don't follow the protocols because alcoholics in withdrawal express symptoms that can be misleading. For instance:

One of the problems is that the condition actually worsens at first. This is because we are depriving them of alcohol, which may lead to hypo-glycaemia. Also, alcohol is a depressant. Withdrawal from alcohol leads to metabolic outpouring. They may become hypokalaemic, with potassium falling along with the blood sugar. This may lead to misdiagnosis by junior staff.[61]

The problem being tackled by Dr Warrington is not unrelated to that described in the textbook, but it isn't the same either. Dr Warrington needs to treat patients, to arrange to have them withdraw from alcohol, and to deal with some of the physiological side-effects of that withdrawal. He also needs to persuade them to abstain from alcohol: 'People have brought alcohol into the hospital, for instance, by injecting fruit. You would be amazed how much alcohol it is possible to inject into a banana.'[62]

Here, then, alcoholic liver disease is being enacted as a set of physiological and anatomical malfunctions. But it is also a regime of treatment, and a matter of psychology too. The similarities with lower-limb atherosclerosis are obvious. Here again treatment co-exists with medical truths. So this assemblage draws on and relates to that witnessed by the textbook, but ramifies off in other directions too – into the practicalities of treating patients, junior medical staff, and nurses.

Just down the corridor from Dr Warrington, Singleton and I interviewed a busy Ward Sister. Sister Fraser is manager of one of the *gastro-enterology wards* in the hospital. This is our third site. Sister Fraser has much experience of patients admitted with the diagnosis of alcoholic liver disease. She and her staff are responsible to Dr Warrington and to the other consultants for the treatment of such patients. She tells us that they are often very poorly when they arrive. In addition, many do not want to accept that they abuse alcohol. The nursing staff start by devising a care plan which includes special nutrition, they check for bed sores, and arrange for the appropriate tests. And then they start on the process of trying to dry the patients out. Alcohol is forbidden, and a sedative, heminevrin, is prescribed to ease withdrawal. Even so, 'the majority are difficult, aggressive because they are withdrawing from alcohol on the ward. Which means that they can be disruptive of ward routine.'[63]

So the process can be distressing for all concerned – patient, nursing staff, and fellow patients. At the same time, the ward staff are trying to arrange appropriate after-care. Here is Sister Fraser:

Some patients have partners who are also alcoholics. We won't be able to help them very much. Socially, we have a social worker who may offer financial advice. Not very many patients get to see the psychiatrist. But we give them information about Alcoholics Anonymous, and also about the Alcohol Information Centre, which offers counselling and support, one-to-one.[64]

However the nursing staff, who often see the same patients readmitted time after time, also worry about the lack of after-care. Here is a colleague of Sister Fraser, who manages the equivalent women's ward:

> I would like to see more support for alcoholics. The fact that there is no psychiatric support makes me mad. Social work support is limited. If they can't rehouse them, can't move them, then they are likely to be going back to the situation which made them drink in the first place. That's distressing. If they want to get out [of the cycle of alcoholism] it would be much easier if they could have proper support.[65]

So the ward is a third site, and those who work on it enact another variant of alcoholic liver disease both as an object in-here, and a context out-there. Thus it becomes: a set of nutritional demands; the administration of drugs; a response in the form of special nursing care; the need for quick responses to occasional dramatic medical events; an ability to manage delusional patients; and, perhaps most interesting in the present context, the process of organising broader community support, for instance from social workers, but also from family, the community health trust, and a range of other agencies and bodies. Here, then, alcoholic liver disease and its treatment becomes some kind of composite. It relates to, and draws upon, a context that is medical, psychiatric, but also social. But both the psychiatric and the social are sources of despair for those who build the disease in this particular way.

And here is a fourth site. For only a few miles from the hospital Singleton and I visited Dr Bowland, a *general practitioner* on the Heathcote estate where there is a substantial problem with alcohol abuse:

> *Interviewer*: Do you talk to patients about the consequences of drinking?
> *Dr Bowland*: This is not an issue. It isn't really possible to talk about the physical consequences of alcohol abuse. I can't talk about such things to many of my clients – to do so might provoke a violent response. The issue is just not relevant to them. They aren't interested in long-term questions, don't take them into consideration. Most people who live in Heathcote have accepted that they will never work again and don't aspire to a fancy car or to different and better housing.[66]

Dr Bowland has been trained in ways that are similar to Dr Warrington and Nurse Fraser. She knows what the textbook says about the effects of alcohol on the liver. This is being crafted as a part of the relevant hinterland. But even so, the world, and therefore the disease, are being enacted very differently. This is because the textbook knowledge and training is crafted together with other quite dissimilar elements to make up a hinterland that is substantially different. Dr Bowland makes the point quite graphically by telling us the story of one client who had been told by a consultant in the hospital that if she carried on drinking she would kill herself. The consultant hoped and expected that this would shock the patient into abstinence. But her reaction was in fact quite different. She simply came to Dr Bowland to ask, almost matter-of-factly, how long she still had to live, how many months. Alcoholic liver disease, Dr Bowland was telling us, was the least of her problems. And to underline the point, she asked us whether we'd seen the drug dealers near the little row of shops close to the surgery. Hard drugs were widely used on the estate. 'Frankly', she added, 'they'd be better off on alcohol.'

So alcoholic liver disease is a very different object here. Arising out of and producing a visible context that is very different in form, it is also being enacted very differently. It is a problem for sure. But for many it is also the least of the various available evils. And in the context of a serious attempt to deliver health care in such underprivileged circumstances to such underprivileged patients, it is rarely a matter of high priority.

Mapping the sites?

I could multiply the sites, but I'll stop with four, the enactment of four versions of alcoholic liver disease and the production of four different contexts. For yes, here again we face the problem of difference or multiplicity. So as with atherosclerosis, we can say that alcoholic liver disease is fractional, that it is more than one and less than many. And then we can look at the strategies for relating these. Many of the items on the list for atherosclerosis are at work here too:

- There is *layering*: a continuing faith, admittedly more relevant in some contexts than in others, that the body contains or is a single reality out there, with particular and quite specific attributes, even if those attributes are not necessarily clear either in the ward or the medical textbook.
- Alongside this there is a *single smooth narrative* – most clearly, perhaps, from Dr Warrington and the textbook, which can account for the various versions and manifestations of alcoholic liver disease by locating it as attributes of, and pathways in, the body, and noting that these express themselves in various visible ways.
- There are *translations* between different 'indicators' of alcoholic liver disease too, though I have not illustrated these in the brief account above. But for instance, there are translations in the hospital for comparing the results of different tests, for converting one to another.
- Then there are *hierarchies* between the tests. In particular contexts some are taken more seriously than others. Dr Warrington, for instance, mentioned that he had introduced a set of new tests for liver function. These are tests, he told us, that have since become the 'gold standard' at certain diagnostic moments, and are better than the older tests.
- There are also *rationalisations* and explanations for apparent discrepancies. For instance, it is a cliché that there are almost always differences between

liver-function test results and the patients' own more optimistic accounts of their level of intake of alcohol. But almost always it is the test results that are treated as the gold standard: those dependent on alcohol are taken to have systematic reasons for concealing the level of their intake.

These are the perspectival strategies that work to produce singularity out there, in the face of difference in here. At the same time, there are also several non-perspectival strategies at work:

- There are sites, and therefore realities, that are *mutually exclusive* in space, time, or both. For instance, some of the tests for cirrhosis described in the textbook are only possible on dead livers, while liver-function tests are only possible on living organs, and patient stories about drinking also come from live people (though not necessarily the person who is drinking).
- Some incompatibilities are simply *kept apart*. This, for instance, is generally the case for the world inhabited and enacted by Dr Bowland, and her clients on the one hand, and the routines of the hospital ward and the practices of consultants such as Dr Warrington on the other. Very frequently these different versions of alcoholic liver disease don't go together, and overlap may, as we have seen, look like tragicomedy.
- Some differences are effaced by being *added together*. What, for instance, counts as an improvement? The hospital says that this is abstinence. Such is one possibility, one reality. Or is it an improved life-style and a reduction in alcohol intake – a second reality? Dr Nixon, a community consultant whom we also interviewed, went for the latter version. But this is not only inconsistent with the general view in the hospital (which would suggest that these realities need to be kept apart) but is also, and in itself, additive (quality of life *and* reduced intake).
- Finally, and related to this, it is possible to create *different objects* – an issue or a strategy of considerable importance to which I will return in the next section.

Obviously many of the strategies for handling difference listed by Mol are also at work here. This means that alcoholic liver disease may be understood as a fractional object. Differently enacted in the different practices in the different sites, those differences are managed in a way that also secures the continued possibility of the singularity of alcoholic liver disease at each particular location. This in turn opens up the possibility of an ontological politics.

One example. As I earlier indicated, many of the professionals believed that after-care was both inadequate and poorly co-ordinated: that far too often, when patients were discharged they were left to fend for themselves – which meant that they returned to their old circumstances, failed, rapidly returned to drinking, and were readmitted to hospital after a few months. This was distressing from both a medical and a human point of view. However, it was also an economic and management problem for the hospital. Patients with alcoholic liver disease kept on returning and occupying costly hospital beds.

But, though the professionals would never put it in these terms, this also implies a version of *ontological politics*. This is because it is not only about organisational failure (though this was one common way of talking about it), but also about the relations between two realities, two different alcoholic liver diseases. There is one, produced by medicine as this is enacted in the textbooks, practised in the wards and in particular in the consulting rooms. This is an alcoholic liver disease that is overwhelmingly *medical* in character. It is located *in the body* and in particular in the liver of the patient. To the extent that it reaches beyond the body of the patient, its relevant visible context out-there, it is likely to relate to the patient's personal or character problems: we were told, for instance, by one consultant that alcoholics are 'devious' because they try to conceal their drinking. The second alcoholic liver disease is partly medical too, but it is less sharply bounded, extending into the psychiatric and (even more) the social relations of the patient. In this version, alcoholic liver disease changes its shape. Yes, it has to do with the body and the liver, but it also reaches into culture and milieu, with the ready availability of alcohol, with the likely ties between alcohol and social life, with life events (stress, loss, and their possible consequence in depression are sometimes said to lead to alcohol abuse), and on occasion with psychiatric illness. It is located *both within and beyond the body*. And its context is similarly diffuse and heterogeneous.[67]

This is ontological politics. Which of these two realities is to be preferred? Or, perhaps more appropriately, how might a satisfactory balance between the two realities be enacted? How should they be related? Indeed, this was precisely the problem that Singleton and I were being asked to think about when we were commissioned to do the study, and it was one that turned out to be quite intractable. It was difficult to handle because of the organisational, professional, and economic distinctions between acute medicine, community medicine, social work, counselling, welfare agencies, and a range of voluntary organisations. An ontological politics, then, here implied an attempt to reorder organisational and professional relations. Or, to turn this round, the particular version of organisational and professional relations in the location we studied produced a powerful and narrowly medical version of alcoholic liver disease – and as a part of this, a strong distinction between the medical on the one hand and the social on the other. The social was being pushed out of the medical context into invisibility. At the same time it also, but much more tenuously and tentatively, enacted the alternative version of the disease in which the medical, psychiatric and social were taken to be interwoven both in the patient and her condition, but also in the context out-there that was taken to be relevant. So there were two objects, and two contexts, one strong, and the other weaker – but the stronger reality with its distinction between medical and social found it difficult to cope with the continual return to drinking by those dried out in the hospital and discharged.[68]

An indefinite object

So much for ontological politics. One of Mol's points – that the social and the (non-social) real are all mixed up in practice – is well illustrated here, as is her insistence on the importance of a praxiography which explores how realities are constituted in practice rather than taken to be fixed. But now I want to return to the strategy of Mol's that I put aside above: *the possibility of creating different objects*. Mol's own example was about the measurement of success of two different interventions. In the study that she cited it turned out that the surgical procedure of PTA (inflating the balloon in the diseased vessel) improved the ankle–arm index, that is the measured pressure of blood to the diseased limb. On the other hand, it made no difference to the onset of inter-mittent claudication: after PTA patients could walk no further without the onset of pain than before. However, the results for walking therapy were the other way round. Patients given such therapy were able to walk further, but there was no discernible improvement in the measurements of their ankle–arm index. So how to deal with this discrepancy? One can imagine various possi-bilities, but in the case she describes the response was to create two different atheroscleroses: walking distance, and ankle–arm index.

We have just come across a similar manoeuvre in the case of alcoholic liver disease. This was the argument put by Dr Nixon, the community health consultant. Most professionals in the hospital believed that improvement demanded total abstinence. But when we put this to him, Dr Nixon didn't agree:

> No. It is not just a question of being substance-free. It also has to do with improving other aspects of life. Such that the substance, or the alcohol, becomes secondary. Then people begin to be free, free of the substance, and enjoy health and a social life. These become more important than the substance. So, for instance, success would be talking with the children a couple of times a week in the evening, instead of going to the pub the whole time.

Dr Nixon is much less interested in abstinence (the equivalent of a perfect ankle–arm index) than he is in enjoying health and a social life (equivalent to the disappearance of intermittent claudication). In effect, then, he has *created a different object*. In his practice, alcoholic liver disease has been turned into something different.

Empirically there is nothing very surprising about this. Remember that the consultants and the nursing staff in the hospital are confronted with patients whose drinking has led to their hospitalisation. In many cases they are gravely ill, and sometimes they are close to death. Dr Nixon's patients have some kind of 'drinking problem' or they would not be talking to him, but they are not being wheeled into Accident and Emergency, and in many cases they are holding down jobs and have partners and children.

So the fact of the multiplication of objects should not surprise us: if Dr

Warrington, Dr Nixon (and the general practitioner Dr Bowland too) are caught up in the enactment of different alcoholic liver diseases and different contexts for those diseases, then this is to be expected. But in the course of the study Singleton and I also discovered something more disconcerting. This was that *our own object of study and its contexts were continually moving about*. Thus we not only found that we were shifting between different alcoholic liver diseases but also, and uneasily, between different problems. Initially we were invited to explore the diagnosis and treatment of *alcoholic liver disease*. Call this object number one. But as we moved into the study and interviewed the professionals we found that we were often talking about *liver disease* (object number two) rather than alcoholic liver disease. Or, more specifically, we were discussing *alcoholic cirrhosis* (number three). Or, very commonly, the talk was of *alcohol abuse* and its implications for individuals and the health care system (four). Or (not necessarily the same thing) it was of *alcoholism* (five). Or (as we have just seen for the case of Dr Nixon) it might be about *overall quality of life* in relation to substance abuse (six). The issue, then, was how to think about this displacement: the fact that the object of study seemed to slip and slide from one interview to the next.

There were moments when we castigated ourselves. We had been invited to consider the system for diagnosing and treating alcoholic liver disease and here we were, shifting between different topics and objects. Indeed, we had promised (in the face of some scepticism, it must be said, on the part of the professionals) to map out that system:

> In the first stage of the research we will seek to map out the processes involved in diagnosing and treating a 'typical' patient with alcoholic liver disease – so to speak, the typical trajectory of a patient within the organisation of medical care.[69]

This being the case, it seemed uncertain whether we should be allowing ourselves to discuss (say) alcoholism or quality of life. Instead we sometimes felt that we should be taking a firmer grip on ourselves and our study, and working to focus it better, to keep it on message. As time went on, however, we came increasingly to the view that this was not only impossible, but might also be counterproductive. We found, for instance, that the different 'maps' for patient trajectories that we derived from our informants in different sites didn't really coincide with one another:

> sometimes patients are referred from the hospital to Castle Street [the Alcohol Advice Centre]. Then they may be given wrong expectations about what can be achieved, and they get lost to the system. In the hospital it depends on who they see. The psychiatric liaison nurses . . . are very experienced, but junior consultants do not have that experience. Wrong expectations are built up, when patients think they can come straight to Castle Street, and do not realise that it is by appointment only.[70]

The hospital map didn't coincide with the Castle Street map. This kind of complaint recurred and recurred:

> Links with GPs are a bit variable. Some don't refer to Castle Street at all. Counsellors will only see people who refer themselves.

Perhaps this was because the health care system was disorganised, but the consequence was that 'mapping' as a metaphor didn't work too well. And trying to trace this in terms of our supposedly major interest, alcoholic liver disease, didn't work either. The reason was that alcoholic liver disease was very often difficult to separate from the other related objects. There might, indeed, be moments when it was possible to separate it from other topics into which we were slipping. Perhaps, for instance, this was possible in the pages of the textbook, or in discussion with a consultant such as Dr Warrington. But predominantly, and in most of the sites, the condition was linked more or less strongly and with greater or lesser specificity, with alternative, partially connected, foci of the kind I have listed above. The consequence was that it became natural to attend to one of the other foci, or some mix of objects, rather than insisting rigidly that talk should focus on alcoholic liver disease itself. For instance, as we have seen, Dr Nixon wanted people to be 'free of the substance'. It wasn't that he was in ignorance of alcoholic liver disease medically understood. It wasn't that he denied its relationship to substance addiction. But the real focus for him was quality of life and this particular version of personal freedom – not enzymatic, histological and anatomical changes in the liver.

We were involved – and participating – in slippage. But how to think about this? As I have just noted, we gradually came to think that this was not simply a sign of shoddy method, a failure to get a grip on something definite. Instead, we slowly came to believe that we were dealing with an object that wasn't fixed, an object that moved and slipped between different practices in different sites. This was an object that, as it moved and slipped, also changed its shape. It was a *shape-changing object that, even more misleadingly, also changed its name.* It was an object whose slippery shape-changing also reflected what the managers and other participants took, perhaps correctly, to be an expression of organisational dislocation, fragmentation and disorganisation. So its relevant context out there changed too.

Shape changing, name changing and fluidity

Does shape changing and name changing necessarily go with disorganisation? Is it something that is to be avoided? Is the methodological difficulty a sign of difficulty in the real world? Does it suggest that objects and their contexts are best when they are fixed and definite? The case of alcoholic liver disease suggests that this might be so. There doesn't seem to be much doubt that the condition would have been better treated and better enacted if the organisation for after-care had been better integrated and more tightly structured, if people in the hospital had understood the remit of the Alcoholic Advice Centre a little more clearly, and if the broader and less bounded version of alcoholic liver disease had been more consistently enacted. Some such recognition is implicit in the words of the consultant who commissioned us to do the study:

> I would like to work more closely and effectively with an alcohol strategy for this district.[71]

But caution is needed. This is because if shape – and name – changing were a problem here, then there may be other locations where this is not the case. For instance, Marianne de Laet and Annemarie Mol describe the development and diffusion of a particular technology, the Zimbabwe bush pump (de Laet and Mol 2000). As they tell it, the pump is simple and robust. It is widely used in the villages of Zimbabwe, where is has often replaced other less satisfactory sources of water that may have been polluted and distant. The pump is manufactured in Harare as a kit, which is then installed after the necessary digging and concreting, by the villagers themselves. The idea – enshrined in government policy – is that villagers should organise themselves into a collective to take responsibility for installing and maintaining the pump. So the pump, it turns out, is not just a matter of technical engineering – and the introduction of pure water – but it is also an instrument of social engineering. Pure water, a simple machine, and a community structure, it is a simultaneous intervention in *nature*, in *technology* and in *society*.

De Laet and Mol are interested in this pump because it is so successful. So why is this? Their argument shows that its success is related to *shape changing*. To describe this they talk of *fluidity*.[72] The pump, they say, is a fluid technology. Thus, though all the pump-kits manufactured in Harare start out life with the same shape, as they are installed and maintained that uniformity tends to disappear. It disappears *technically*. As bits wear out or fall off they are replaced by local make-do. These, it turns out, often work just as well as the original broken part. Even the inventor of the pump, who travels to the villages from time to time to see what is happening to the pumps after they are installed, is sometimes surprised by the innovations devised by villagers to keep their pumps going. So uniformity disappears technically. But it also disappears *socially*. For instance, though government policy insists on the creation of a village collective to take responsibility for the pump, very often this doesn't happen, and the pump is cared for by a few families, or in some other kind of shared arrangement. Finally, *nature* (if I may extend the definition of nature to include what will count as pure water) also turns out to be variable. Is the water from the pumps pure? The answer is, almost always. But what does this mean? Sometimes it means that the water is pure because it meets bacteriological standards: the E.Coli count is lower than the internationally agreed threshold. More often, however, in a country where most of the villages are far from the facilities of a bacteriological laboratory, it means that few villagers contract water-borne diseases. (And there are instances where villagers are healthy even though the water fails the bacteriological tests.) 'Pure water' is also a variable commodity.

So why is the pump so successful? As I have noted above, de Laet and Mol are telling us that this is because it changes shape. It is, they say, a fluid technology. And it is its fluidity, the fact that it slips and slides from one village to the next, that explains the extent of its diffusion. It slips, slides, alters its shape – and its relevant context alters too. Indeed the implication is even stronger than this. It is that if the technology were rigid – if the manufacturer insisted that repairs be carried out using parts made in the factory in Harare, or the government enforced its policy about village collectives, or international bacteriological standards were the sole measure of water purity – then the pump would not be nearly as successful as it actually is. In short, it is fluidity, the capacity for shape changing and remaking its context, that is the key to its success.

The conclusion is that fluidity, shape changing, and indeed name changing are not problems in and of themselves. They *may*, of course, be problems. Perhaps this is the case for alcoholic liver disease in Waterside, though even this is not entirely clear. This is because the problem may be not that the object changes its shape and its name, but rather that the balance and the movement between the different objects and their contexts is unsatisfactory: because it isn't sufficiently fluid. Thus the body-based and predominantly medically produced version of alcoholic liver disease in the consulting room and in the medical functions of the ward is surely appropriate under some circumstances. If a patient is dying, then what is required is a strong medical regime and a strong medical reality. But, if the patient has withdrawn from alcohol, then what is needed is more likely to be the alternative broader reality produced by a mix of medicine, psychiatry and social support. Which (in the name shifting I have discussed) has as much to do with alcoholism or substance abuse as alcoholic liver disease. The issue, then, is about the relations between different objects and their different contexts. A graphic way of making the point would be to say that the consultants and others caught up in the narrowly medical assemblage 'ought' to be much more interested in the broader medical-psychiatric-social reality of alcoholism – and the assemblage that crafts this than they actually are. And they 'ought' to be much more interested in how patients are shifted from the acute context of alcoholic poisoning in the ward to the process of support in the community. For that process, the displacement from one method assemblage to the next, 'ought' to be much easier, much more fluid, than it actually is.

Definite fluidities?

If real objects and their visible contexts are enacted in practices, then there are issues about how those practices relate together. This is the problem of difference, or the problem of multiplicity. And there are various metaphors for handling this and dealing with bundles of partially overlapping methods assemblages. As we have seen, Mol describes a series of strategies for holding singularity and multiplicity together. Perhaps she can be read as suggesting that this oscillation between the single and the multiple is a chronic condition of being in Euro-America: that the assumption of singular out-thereness is an enactment that sets limits to the Euro-American conditions of possibility, even if it is also, and in some senses, misleading. Misleading because the method assemblage depends on, grows out of, and is enacted by mechanisms of inter-ference between practices which depend on separation while also insisting that they are joined. As we have seen, there are other arguments similar to this. Thus in Latour's understanding modernity grows out of an analogous distinc-tion between pure forms, pure distinctions on the one hand, and a proliferation of heterogeneities, impurities and hybrids on the other. His argument is that *both* heterogeneities or impurities *and* attempts at purity and distinctness are necessary: it is the pretence of the latter that allows the fecund but more or less concealed production of hybrids.[73]

Perhaps this is so, but there are further questions arising. For instance, multiplicity also poses questions about *definiteness*, the last of the Euro-American versions of out-thereness detailed in Chapter 2. Is it a sin to be indefinite? Perhaps. But as we have seen, out-thereness and its in-here objects may look indefinite because they are slippery, changing their shape and perhaps even their name. But if this is right then perhaps we also need to change our sense of what it is to be definite. Perhaps we need to say that the shape shifters and the name changers *are* indeed definite – but that they also change their shapes and names. To do this is to redraw the boundary between two parts of out-thereness – between that which is visible and that which is not. It re-works the conditions of possibility for Euro-American method. It makes it quieter and more generous.

But there is a more radical possibility too. We might, instead or as well, imagine versions of method assemblage that craft, sensitise us to, and apprehend the indefinite or the non-coherent in-here and out-there. This is the challenge that I address in the next chapter.

INTERLUDE:
Notes on presence and absence

In the 1920s Ferdinand de Saussure wrote that: 'The linguistic sign unites, not a thing and a name, but a concept and a sound image' (Saussure 1960, 66). In his synchronic linguistics the sign thus links sound image (signified) and concept (signifier). The relationship between signifier and signified – the sign – is arbitrary, and the value of the sign also depends on its difference from other signs: 'child', 'woman', 'man', these terms achieve their value relationally.

This relationality has been the working tool for structuralist and post-structuralist writers. Structuralists hoped and searched for a fixed syntax of relations that reflects mental processes and actions in the world. Post-structuralists abandoned this search for foundations. There are always things that cannot be told, that cannot be made present. Instead they explored limits and incompletenesses. For Michel Foucault the limits to the conditions of possibility were or are set by the (modern) episteme. For Jacques Derrida the traces of incompleteness can always be discerned in the erasures and aporias enacted in whatever is present: in the deferrals of *différance*. Nothing is self-sealing, complete. Not everything can be known: it depends on what is not there. The argument is against what these philosophers call a 'metaphysics of presence': the idea that everything could be brought together and created or joined or known in a single location. *What is being made present always depends on what is also being made absent.*

The relationality of synchronic linguistics and its post-structuralist successors has also been a working tool in the sociology and philosophy of science and technology. For instance, Donna Haraway's arguments about partial connec-tions, cyborgs and situated knowledges are composed in a different idiom, but their overall shape is similar. Collapse to unity is never a possibility, even though it claims to be. In addition, relationality underpins the work of Latour and Woolgar, and of Mol. Writers such as Foucault and Derrida insist on the materiality of relations and of the trace.[74] Foucault's understanding of 'discourse' reveals its thoroughgoing materiality. Feminist theory has explored the enactments of embodiment, and the materialisation of (for instance) heterosexuality and the displaced traces of other possible sexualities.[75] Prodded by feminist sensibilities, science technology and society (STS) has come more recently to embodiment (for instance in the writing of Charis Cussins, Tiago Moreira and Ingunn Moser, as well as that of Haraway and Mol).[76] Nevertheless, it is perhaps in STS that materiality has been most attended to. Thus we have seen the way in which Latour and Woolgar consider the relations of science both in the form of traces and statements and in the shape of other materials such as inscription devices. This is a thoroughgoing *relational materiality*. Materials – and so realities – are treated as relational products. They do not exist in and of themselves.[77] And the same logic is used to explore, in particular, the relations between materials and statements. We have seen that these are pictured as more or less precarious chains of relations. The links in these chains then get deleted, pushed into invisibility out-there, in the final product when suddenly all the intermediate steps are made to disappear, and we are confronted on the one hand by a visible fact out-there, and on the other hand by a statement in-here that describes that reality and which appears to derive from it. Latour and Woolgar can be understood, then, as calling for the rehabilitation of the mediating relations which produce statements and visible realities. For the rehabilitation of the necessary but invisible work which produces these.

This is a version of method assemblage. It is the making of relations. So what do those relations do? What is distinctive about them? What *is* method assemblage? I have offered two provisional responses. In Chapter 2 I defined this for the case of representation, as the enactment of a bundle of ramifying relations that shapes, mediates and separates representations in-here, represented realities out-there, and invisible out-there relations, processes and contexts necessary to in-here. In Chapter 3, I offered a parallel definition appropriate to objects: that method assemblage is also the crafting of relations that shape, mediate and separate an object in-here, its relevant context out-there, and then an endless set of out-there relations, processes and all the rest that are a necessary part of the assemblage but at the same time have disappeared from it. As is obvious, the two are similar in form. But the post-structuralist philosophical tradition suggests a different vocabulary. If we use this then method assemblage becomes the enactment of *presence, manifest absence*, and *absence as Otherness*. More specifically, method assemblage becomes the crafting or bundling of relations or hinterland into three parts: (a) whatever is in-here or *present;* (b) whatever is absent but is also *manifest in its absence;* and (c) whatever is absent but is *Other* because, while it is necessary to presence, it is not or cannot be made manifest. Note that it is the emphasis on presence that distinguishes method from any other form of assemblage. Note also that to talk of crafting is not necessarily to imply human agency and skill. The various ethnographies we have explored suggest that people, machines, traces, resources of all kinds – and we might in other contexts extend the list to include spirits or angels or muses – are all involved in the process of crafting.

A comment, now, on presence, manifest absence and Otherness.

Presence is, obviously, what is made present or (as I shall sometimes say) *condensed* 'in-here'. Latour and Woolgar talk of statements: these are versions of presence. Mol talks about angiographies, but also atherosclerotic blood vessels: these are further versions of presence. Others that we have touched upon include technical objects such as bush-pumps, and such shape-changing entities as alcoholic liver disease. All of these are presences enacted into being within practices. Some are representations while others are objects or processes. Presence, then, is any kind of in-here enactment.

Manifest absence goes with presence. It is one of its correlates since presence is incomplete and depends on absence. To make present is also to make absent. Examples. The statements in the papers of the Salk Laboratory describe endocrinological realities. Present angiographies describe diseased blood vessels. Excised blood vessels grow out of or have implications for future regimes of treatment, or past disease-inducing events. Alcohol abuse on a sink estate makes and depends on a context quite unlike that generated by alcoholic liver disease in a hospital gastro-enterology ward. The context of a bush-pump in a village is not much like its context in, say, its original manufacture. Each of these is a presence and a manifest absence, in one form or another.

Otherness, or absence that is not made manifest, also goes with presence. It too is necessary to presence. But it disappears. Perhaps it disappears because it is not interesting while it goes on routinely (the power-supply to the Salk Laboratory? The pay cheques to hospital staff? The current organisation of health care? A factory in Harare with a non-proprietorial proprietor?). Perhaps it disappears because it is not interesting, full stop (the availability of alcohol, a pub-going culture, a broken marriage). Perhaps (though no doubt this is an overlapping category) it disappears because what is being brought to presence and manifest absence cannot be sustained unless it is Othered (the social-and-cultural context for alcoholic liver disease in the context of the consultant's office? The active character of authorship in the production of Salk Laboratory statements?). The implication is that Otherness takes a variety of forms. Those above – *routine, insignificance* and *repression* – are no doubt only three of the possibilities.[78]

It follows that method assemblage is also about the crafting and enacting of *boundaries* between presence, manifest absence and Otherness. These boundaries are necessary. Each category depends on the others, so it is not that they can be avoided. To put it differently, there will always be Othering. What is brought to presence – or manifest absence – is always limited, always potentially contestable. How it might be crafted is endlessly uncertain, endlessly revisable. Normative methods try to define and police boundary relations in ways that are tight and hold steady. An inquiry into slow method suggests that we might imagine more flexible boundaries, and different forms of presence and manifest absence. Other possibilities can be imagined, for instance if we attend to non-coherence. It is in this spirit that in the next chapter I consider allegories and events as versions of method.

5 Elusive objects

Battles do have causes and consequences, and lead to subtle reflections when they are considered years later. But none of that is available to those who take part in them. They and their comrades – and their enemies – are sealed into an experience which has no context and no comparison, a present consisting of jokes, terror, trained reactions, insane orders from above, utter exhaustion, the taste of stewed tea, the sound of incoming mortars.

(Ascherson 2002, 15)

That which is not said

During our work on alcoholic liver disease, Vicky Singleton and I visited the Alcohol Advice Centre at Castle Street that I have mentioned above. In one of our papers we jointly wrote as follows (again I am grateful to Vicky Singleton for allowing me to make use of this joint work here):

Finding the door is difficult enough. In a terrace, between two cheap store-fronts in a run-down part of Waterside. The kind of street only three blocks from the big store that doesn't make it. That doesn't make it at all. That smells of poverty. That speaks of hopelessness.

It is a nondescript door. Unwelcoming. A tiny spy glass. An inconspicuous notice. Nothing very obvious. Nothing very appealing. We are ringing the door-bell. Is anyone listening? Has anyone heard? Dimly we hear the sound of footsteps. We sense that we are being looked at through the spy glass. Checking us out. And then the door opens. And we're being welcomed through the door by a middle-aged woman. To find that there isn't a proper lobby. Instead, we're facing a flight of stairs. Carpeted, cheaply. Yes, shoddily.

So we've been admitted. We are, yes, Vicky Singleton and John Law from Lancaster University. And now, we're being led up a flight of stairs. And the building is starting to make an impression. An impression of make-do. Of scarce resources. Of inadequacy. For we're being told people have to come up all those flights of stairs. Some of them can hardly walk through drink. And some can hardly walk, full stop. Up this long flight of stairs. For we're in the kind of Victorian building where the rooms on the ground floor are twelve feet high. Big fancy three-storey houses. Built at a time of optimism. At a time of some kind of prosperity. Which, however, has now drained away.

So the clients need to negotiate these stairs, turn around the half landing, up a further short flight, and then they are on the first floor. Next to the room that is the general office, library, meeting room, leaflet dispensary, the place with the filing cabinets, the tables, the chairs. People are milling about. At the moment no clients, but a researcher who is smoking. Several social workers, the manager, community psychiatric nurses coming and going.

The leaflets and the papers are spilling over everything. Brown cardboard boxes. Half-drunk mugs of coffee. New mugs of coffee for us. Clearing a bit of space. Not too much. There isn't too much space. Files and pamphlets are pushed to one side. Two more chairs. And the numbers in the room keep on changing as clients arrive, or people go out on call, or the phone rings. One client hasn't turned up. Relief at this. The pressure is so great. And then there's another with alcohol on his breath. A bad sign.

The staff are so keen to talk. Keen to tell us about their work. Keen to talk about its frustrations and its complexities.

(Law and Singleton 2003)

What to make of this? Singleton and I argued that this scene could be understood as a manifestation or an expression of the parlous state of after-care for patients. We argued that organisational fragmentation and shortage of resources were reflected in this run-down building and the events that went on in it. It was, for instance, inappropriate to house an alcohol advice centre up a long flight of stairs. It worked badly that there was no proper meeting room. It was also strange to discover that those working on the premises were employed by several different organisations with different conditions of work. The chaos of leaflets – from twenty or more different sources – also reflected, in a manner concentrated on a single set of shelves, the criss-crossing plethora of locations, organisations, facilities, and policies that were all, somehow or other, more or less relevant to the issue of alcohol abuse in the district, yet didn't quite fit together either. We argued, in short, that the building reflected, witnessed, or condensed a wider state of disorganisation. And, as a part of this, it also enacted the interrupted flow between the tighter biomedical realities of the consultants in the hospital and their less bounded, psychiatric and social (but also medical) alternatives being crafted elsewhere.

But we also argued that there was something important about the scene that *could not be put into words* and escaped the possibilities of language. At the same time we proposed that this resistance to explicit formulation was not necessarily a problem. Indeed, on the contrary, we suggested that it might be perfectly appropriate to imagine representation in ways that wholly or partially resisted explicit symbolisation. In short, though we did not put it in these terms, in effect we were arguing that the disorganisation out-there (manifest absence) was being brought to presence and enacted by the premises themselves, or by our verbal but also emotional and aesthetic interaction with those premises. Instead of being Othered.

Allegory

Allegory is the art of meaning something other and more than what is being said. Closely related to irony, and also to metaphor, it is the art of decoding that meaning, reading between the literal lines to understand what is actually being depicted.

It is sometimes said that allegory is a lost art form and that in Euro-America we have lost the craft of saying or representing things indirectly. If this is right then perhaps it is for two reasons. The first is that allegory flourishes as an art form in contexts where there is explicit repression. If the regime (or the church, or the elders) do not tolerate criticism, then the conditions are in place for allegory. It has, then, been cultivated as an explicit art at certain periods in European history. Audiences have gone to the theatre to see a play (say) about classical Rome knowing full well that while they might be learning something about Rome, they were also learning about those currently in power. At present Euro-Americans are mostly lucky enough to live in countries which minimise overt political repression. To that extent stylised forms of allegory are not cultivated as a necessary form in public life.

In the present context the second reason is more interesting. This has to do with the dominance of literal representation. Descriptions describe directly. This is the goal, and seemingly the achievement, of many or most of the major forms of representation in Euro-America. Physical science, biomedicine, social science, but also politics, journalism and current affairs: method is assembled in all of these in the form of statements (or other representations) that correspond to manifest absences in straightforward ways. What is made manifest is, so to speak, constructed as being straightforward. More particularly, as the ethnography described in Chapter 2 suggests, the reality made manifest is said to author(ise) the representation that seemingly derives from it.

As an argument for the disappearance of allegory this sounds plausible. Direct representation is indeed celebrated. At the same time, the argument does not quite work. First, as I also argued in Chapter 2, direct representation is never direct. It is mediated. If a statement in endocrinology (or medical sociology) corresponds to a reality out-there, if it simply seems to describe it, then this is because most of the assemblage within which it is located has been rendered invisible, Othered. The authorship, the uncertainties, the enactment of out-thereness, all of these have disappeared. The *appearance* of direct representation is the effect of a process of artful deletion. So the argument we need to make is this. On the one hand, indeed it is the case that direct representation offers no overt space for allegory.[79] But on the other hand direct representation is *built* in allegory. There is nothing direct or literal about the link between present statements and the absent realities. This means that those statements come out (or are telling) of something other or more than the reality they describe. They are effects of allegory that conceal their allegorical origins. That is what representation is: *allegory that denies its character as allegory.*

The argument, then, is that wherever there is depiction, so too, there is allegory. So it is not that allegory has been lost, but rather that it is covertly practised. Or, to put it differently, we are all steeped in the art of allegory. Natural scientists, social scientists, politicians, journalists, workers by hand and by brain, all of us are expert allegorists. All of us are skilled in reading between the lines being fed to us. All of us are consummately skilled at saying what we mean rather than what we are saying. Politicians, advertisers, lawyers, satirists, liars, diplomats, conciliators, priests, parents, partners, general practitioners – all of us trade in allegory, and all of us are skilled in the practices of decoding it.

But there is a further twist. As I have noted, an overt commitment to allegory flourishes in circumstances of overt oppression. But if we rephrase this slightly, we find that allegory flourishes in circumstances of contested authority. It is something like this. The powerful (try to) insist that their statements are literal depictions of a single reality. 'It really is that way', they tell us. 'There is no alternative.' But those on the receiving end of such homilies learn to read them allegorically. Cynicism, scepticism, the detection of hidden interests, a sense of the ideological, these are the techniques used by subordinates to read through the words of the powerful to the concealed realities that have produced them.[80]

Apply this lesson back to the representational regimes of natural and social science. At first sight these are authorities, sources of authorised knowledge. Environmentalists know about the environment. Demographers know about populations. But, at the same time, the realities so made are also contested. Variably, to be sure. Perhaps sociology, and even worse cultural studies, carry little weight in certain locations.[81] But for economics, demography, and many parts of natural science this is not the case.[82] As institutions of authority they (try to) insist that their statements are literal depictions of a reality thereby made manifest. 'Reality is that way', they tell us, at least within technical restrictions. But here too allegorical readings are common. Experts are not trusted.[83]

There is a substantial social science literature on experts and trust. The argument comes in a number of guises, but one is about the 'public understanding of science'. High-status scientists worry about the scepticism shown by lay people about natural science, and there have been several publicly funded research programmes on this in the UK over the last twenty years.[84] Those in authority believe that lay people do not properly appreciate and respect science and its methods – and that this is something that should be put right. But to think this way is asymmetrical. It assumes that scientists with their methods know best. That the realities they describe and make are prior, independent, and all the rest. It is, in short, to refuse the legitimacy of lay and allegorical readings of science. Reading at least parts of science and its claims allegorically rather than literally, lay people are asking (for instance) about the special interests that lie behind claims about the truth. They wonder, for instance, exactly why a Minister of Agriculture appears on television feeding a hamburger to his obviously reluctant daughter.[85] And they fear the worst.

STS scholar Brian Wynne has written about the lack of trust in experts. Following the Chernobyl disaster, for instance, he watched the interactions between hill farmers in Cumbria and the experts who came to monitor the fallout on the Lakeland hills and check on the safety of the lambs to enter the human food chain. There are various themes in his work, but one of them has to do with scepticism, with lack of trust. Why should we believe the experts, the hill farmers were saying or thinking, when they are paid by government agencies that appear to be linked to the nuclear industry? Why should we believe them when they come on to our hills in the wake of the Chernobyl disaster and tell us how to treat our lambs in ways that we know from years of farming experience make absolutely no sense at all? Why should we believe them when it turns out that their confident predictions about the binding of radioactive caesium in the hill fields are simply wrong, and that months later the sheep still depend on imported fodder?[86] Wynne's account shows that people like hill farmers are working allegorically. That they are moving the boundary between what is manifest and what is Othered around. Official versions of the manifest, the literal accounts offered by experts, are being doubted. Parts of what is Othered in those versions are being brought into view, made real.

And this is what allegory always does. It uses what is present as a resource to mess about with absence. It makes manifest what is otherwise invisible. It extends the fields of visibility, and crafts new realities out-there. And at least sometimes, it also does something that is even more artful. This is because *it makes space for ambivalence and ambiguity*. In allegory, the realities made manifest do not necessarily have to fit together.

Ambiguity and ambivalence

So this is the art of allegory: it is to hold two or more things together that do not necessarily cohere. Wynne, writing in a different idiom but making a related point, says that:

> Through their rationalist discourses, modern expert institutions and their 'natural' cultural responses to risks in the idiom of scientific risk manage-ment, tacitly and furtively impose prescriptive models of the human and the social upon lay people, and these are implicitly found wanting in human terms.
>
> (Wynne 1996, 57)

His argument is that lay people both fall into line with these models *and* that they don't. We might add that they both enact those models and they don't. This, then, is the possibility opened up. Low-status people need to live in and enact two versions of reality – the official and the local – at the same time. Such is Wynne's argument.

Vicky Singleton has made similar arguments about ambivalence. In her work she has explored the way in which health-care programmes often, perhaps usually, represent their aims and their successes in ways that are clear, singular, definite, and are said to follow from explicit policy. At least in public those responsible for the programmes usually craft a representation and a reality that is singular and definite. They say, for instance, that the UK cervical screening programme has successfully reduced women's deaths from cervical cancer. However, Singleton also shows that at the same time there are other stories that represent and enact the programme in ways that are less definite, less coherent, more dispersed, and have relatively little to do with the direct application of policy. In other words, her argument is that ambivalence runs through the cervical screening programme.

For instance, it turns out that general practitioners simultaneously entertain doubts about the programme, and believe that it is good for women to be screened. Here are the interview words of a GP:

> We do persuade women to have smears and, but [pause] . . . the evidence that smears have done a lot of good is not great in this country [pause] . . . I don't have any difficulty encouraging women to have a smear but I wouldn't, pressure is something I would not like to exert on a woman.
>
> (Singleton and Michael 1993, 258)

This is an ambivalence in words. The GP is awkwardly hinting at – and recognising – two realities. On the one hand there is a programme that reduces rates of the disease, and on the other hand there is one that does not. But such ambiguities and multiple realities abound. For instance, laboratory cytology is also said to be reliable, while at the same time it is taken to be uncertain.[87]

> Now this is where some people can fall down because with the more florid cancers you may not get a positive result. There are ones that are missed and they were staring you in the face, because the smear, which wasn't reliable under this technical set of circumstances, was negative. So the naked eye appearance is important.
>
> (Singleton and Michael 1993, 243–244)

And these are just two of the ambiguities. Writing with Mike Michael, she argues that the screening programme 'is rendered durable by the way that actors at once occupy the margins and the core, are the most outspoken critics and the most ardent stalwarts, are simultaneously insiders and outsiders' (Singleton and Michael 1993, 232).

The argument is not that these oscillations and ambivalences are signs of bad faith. Perhaps there is sometimes bad faith, but something much more important is going on. Rather, Singleton is telling us that programmes such as these *always* harbour and enact conflicting practices and views. And then she is making a stronger claim too. This is that it is such oscillatory ambivalence that makes them possible and more or less successful in the first place.[88] Consistency or coherence can only be achieved in theory and not in practice. Or consistency depends on non-coherence.

GPs, hill farmers, all of us are allegorists because we read between the lines and manifest realities that are not being spoken about in as many words. We play, that is, with the boundaries between that which is Othered and that which is manifest. We might thus think of allegory as a mode of discovery – so long as we understand that in a world of enactment, allegory is also *crafting* what it is discovering. That is the first point. The second is equally interesting. It is that as allegorists, a lot of the time we are crafting and manifesting realities that are non-coherent. That are difficult to fit together into a single smooth reality.

We have been here already, but in a different vocabulary, for this is a point about multiplicity. The different realities enacted in the different practices of the cervical screening programme are indeed different. They are like the lower-limb atheroscleroses discussed by Mol. And again, like the lower-limb atheroscleroses, there are various ways in which these can be – and are – patched together in practice. The possibilities enumerated by Mol apply again here. But having said this, Singleton's allegorical concern is also slightly different. The interest is not simply that presences and absences are multiple. It is also on the *representation* of non-coherence or multiplicity. This is a touchy issue because, as we have seen, Euro-American assumptions about what is out-there make it difficult to think of this or talk of it as non-coherent or multiple. Within the technical limits available, and subject always to the need for correction, it prefers to represent manifest reality as singular. The consequence is that it tries to deny the possibilities for non-coherent depiction offered by allegory. It tries to draw a firm line between those absences that are permitted to manifest themselves and those that do not fit, those that are Othered.

No doubt there are many possible ways for enacting allegory, many possible ways in which it might mess with absence, recrafting the boundaries between realities made manifest and realities Othered. There are many possible ways in which it might stand for and enact non-coherence out-there. Sometimes it will use words – in my reading this is what is being attempted by Singleton and Wynne in the examples discussed above. Sometimes, however, it will take us outside words. And this, I suggest, is what is happening in the case of the run-down building occupied by the Waterside Alcohol Advice Centre. The building – and our apprehension of the building – are an exercise in allegory. In the absence of words I guess that there is less pressure to narrative consistency. There is less pressure to manifest an absence that is single and coherent. Perhaps, then, architectures and other non-linguistic verbal forms are rich sources for allegory. Perhaps they *are* allegories which enact the non-coherent, allowing us to make it manifest. Perhaps it is simply that we are not very good at treating them as allegories – apprehending the ways in which they craft and relate sets of realities that cannot be located in a single narrative.[89]

Of course non-coherence is not a good in itself. As I have noted, Singleton and I were persuaded that a higher degree of organisational coherence would have been better all round for the treatment of those with alcohol-related problems in the Waterside area. Nevertheless, to try to shoehorn non-coherent realities into singularity by insisting on direct representation and Othering whatever does not fit is also to miss the point. It is to (try to) enact a particular version of ontological politics. And it is the strength of an allegorical attitude to method assemblage that it does not miss that point. That it carries an alternative politics. *That it softens and plays with the boundaries between what is Othered and what is made manifest.* That it discovers – and enacts – new and only partially connected realities.

I now extend this argument by exploring a further case, that of a calamitous railway accident.

Ladbroke Grove

On 5 October 1999, a three-carriage Thames Train diesel unit ('165') collided with a First Great Western High Speed Train ('HST') at Ladbroke Grove, two miles outside London's Paddington railway station. The result was devastation.

> It has been established, (reported a barrister, opening the subsequent public inquiry), that a total of some 575 people were travelling in the trains. 31 people died in the crash or from the injuries sustained in it. 23 of the dead were passengers on the Thames Trains' 165, 6 were passengers in the High Speed Train. In addition, the drivers of both trains were killed. Approximately 414 were injured, many very seriously. It follows that over 75 per cent of the passengers either lost their lives or were injured to a greater or lesser degree. The figures for the 165 are even more stark. The best estimate is that it was carrying some 148 people. Of those 23 died and 116 were injured. Only 6 emerged unscathed. 227 were taken from the scene to hospital. Many are continuing to suffer from their injuries and from the shock of exposure to scenes of horror and of devastation.[90]

The trains collided virtually head-on at a closing speed of about 145 miles an hour, and the destruction was horrific. The leading power car and the two leading coaches of the High Speed Train together with the leading two coaches of the 165 Thames Train were very severely damaged. Though the greater damage was to the carriages of the less heavily built Thames Train, the effects of the collision were compounded by the outbreak of a ferocious fire, in part caused by escaping diesel fuel, which completely destroyed the interior of one of the coaches of the High Speed Train.[91] With exit and communication doors blocked, many passengers found that they were unable to escape from the wreckage. They were caught up in the fire and burned to death, or sustained terrible and disfiguring injuries. The extent to which the rescue services were able to help was similarly limited: nothing could be done to extract those caught in the most severely burning coach, and nothing could be done to extinguish the inferno which reduced the contents of that coach to a fine ash.

The Ladbroke Grove accident led to a crisis for the British rail system. There was a widespread belief that things had gone horribly wrong, not just at Ladbroke Grove itself, but also, and much more generally, for the railways as a whole. The result was a public inquiry established to explore not only the proximate causes of the accident, but also background factors. That inquiry (the quotation above comes from the transcript of its proceedings) reviewed the evidence, collected statements from, and cross-examined hundreds of witnesses, and reported its findings nearly two years later.[92] The implications of the collision (together with a number of other railway disasters) were to lead, in due course, to a major reorganisation of the UK rail network.

Collision as allegory

The Ladbroke Grove collision can be understood as a bundle of relations and entities. Some are brought to presence in the terrible scene of the accident itself. Others are made manifest as in the form of a context, relevant to that presence in one way or another. Yet others are rendered invisible, Othered. So the accident is an object. But, like the alcohol advice centre, the scene may also be understood as allegory. This is because, while it does not take the form of statements about reality, if we read between the lines of the carnage it indirectly depicts, enacts and manifests a range of realities. And indeed, this is why I want to go into it. Unlike a representation, what is made present does not pretend to speak for itself. It calls for interpretation. It presses us into allegory. At the same time, it allows us to explore the anatomy of allegorical investigation.

But there are two ways of doing this, of treating it as allegory. The first is to go into it in the form of words and make a consistent linguistic account. This is what happened at the public inquiry, and in the report issued at the end of that inquiry. The report crafts and represents a reality in the form of the circumstances that led to the accident. The second is to try to apprehend the wreckage and the horror without attempting to build a single discursive account. Both are allegorical strategies. Both are possible. Indeed both are important. But it is obvious they work in different ways. First, then, the inquiry and its report.

This is a meticulous investigation into the relations bundled together and brought to presence in the collision. It is the meticulous depiction – and enactment – of a set of relations, a reality that led to the collision. In practice it starts by showing that the driver of the 165 Thames Train made a mistake.

He drove the train through a signal, SN 109, that was set to danger, red. But why? The possibility of suicide is explored and discounted. Neither is it recklessness: the driver had, it is established, a fine record of defensive driving. Rather it is his inexperience that is said to be the problem. He had recently completed his training, and was not particularly familiar with the complex routes out of Paddington Station. But why? The investigation goes on to explore the character, many would say the failings, of the training of drivers at Thames Trains. But what was the problem with the training? Again there are various answers, and some are quite indirect. Drivers may no longer have long-term experience. Previously promoted after a long career on the railway – and experience to match – now many drivers were being appointed through recruitment campaigns and intensive training packages.

But this is merely one part of the relevant reality created by the inquiry. Another has to do with the signal, SN 109. Why did the driver pass this signal when it was red? Having ruled out suicide, the investigators looked to see why he might have overlooked or misread the signal. Again there are various branches and bundles here. Perhaps the innocent misuse of a safety device, the driver reminder appliance, had reassured him that the signal was really set at green. (Back to training.) But why had he not seen that it was red? Perhaps he was distracted. This could neither be proved nor disproved. Perhaps, however, the collision also enacted the angle of the rising sun. (Arguably, this was being reflected by the signals back into the eyes of drivers, and might have given the appearance of a green light.) Or perhaps it was a sign of some fundamental flaw in SN 109 itself. For instance, perhaps what was important was the fact that it was one of a number of signals attached to a single gantry spanning a number of winding tracks with relatively limited lines of sight for train drivers as they approached. This meant that train drivers often needed to count along the line of signals on the gantry to decide which applied to their train. Or (to move on) perhaps it crafted a number of failures by Railtrack, the company responsible for track and signalling. Perhaps, in particular, it indexed the failure of the company to investigate a number of previous incidents at SN 109, and put right a signalling arrangement that was (clearly?) less than satisfactory. In which case perhaps it also condensed the failure of Railtrack to find an effective organisational arrangement for tracking down and seeking to remedy incidents due to faulty readings of signals. Or perhaps it retracted the dangers of a railway organisation in which the track and signalling belonged to one private company, and the trains running on the track belonged to other quite different companies. Or where track and signal maintenance was contracted, and then subcontracted out to a plethora of other commercial organisations, some of which had few railway-experienced staff. In which case it also articulated and enacted, if not the privatisation of the railway itself, then at least the manner in which that privatisation was achieved, requiring, as it did, the fragmentation of a single company, British Railways, which had previously owned and run the entire system, in order to achieve the benefits of market competition. Which (in this version) generated a degree of non-coherence through the

British rail system that makes the alcoholic liver disease case look like a model of good practice.

The inquiry and its report is impressive. It lists and explores a very large number of possibly contributory causes. In debate and in cross-examination in the quasi-judicial public hearings, possibilities are entertained, examined and assessed – and in the report they are accepted or dismissed. The specific details of the process are not those of the Salk Laboratory, but the overall character of the framing is the same. It is to craft statements describing a reality that will stand up in a network of other statements, materials and practices. That reality should be a coherent account, a meticulous enumeration of direct and contributory causes that combine together to produce the accident. Note that this implies the need for endless determinations about the location of the boundary between what is real and to be made manifest on the one hand, and what is to be Othered on the other. Some of those determinations are explicit. They take the form of negotiations, modalisations, and demodalisations. Is this particular description *really* what happened or not? Is it plausible? Is it motivated by hidden interests? Should it be taken seriously or can it be dismissed?[93] Other determinations are less overt. In particular, reality is taken to be definite, singular, prior, and independent – and is made that way. For, yes, the framing assumptions of Euro-American metaphysics are still hard at work. A coherent account of the world is possible even at moments when things have gone dreadfully wrong. The inquiry – necessarily allegorical because nothing speaks for itself, nothing is transparent, everything has to be read as a symptom, as being about something else – thus denies the possibility of non-coherence, multiplicity, priority, and all the rest.

In the real conditions of the inquiry there is little choice. It was charged by the UK statutory body responsible for industrial and workplace safety:

> To inquire into and draw lessons from the accident . . . taking account of the findings of the HSE's investigations into immediate causes.[94]

The requirement for coherence, then, is built into the conditions of possibility. A single report is required, and in one way or another a single reality will necessarily emerge eventually. It will be multi-factorial. There will be many contributory causes to the accident. But they will be drawn together and mapped. The railway reality will, so to speak, cohere in its incoherence. The accident was caused by a determinate set of circumstances. The issue is to determine their character.

So that is one possibility. But what of the accident itself? What happens if we treat this more directly as allegory? This is the alternative option. To treat it as a moment, a dreadful enactment of presence written not in texts and statements, but in steel and flesh and fuel and fire. Written as impact, collapse, inferno, agonising pain, terrible burns, grief, panic, and death. If we do this, then the collision may be understood as an inscription device that writes its texts not with pens on sheets of paper, but rather with the instruments of kinetic force and fire upon the bodies of people, and the twisted wreckage of gutted rolling stock. That inscribes itself in and through a theatre of cruelty.

This is the stuff of nightmares. But at the same time, horrific though it is, it is not entirely inappropriate. If statements have to take the form of direct representations, then the collision does not craft direct representations. But it does produce something like statements that can be apprehended as being about something else – so long as we are willing to think allegorically and move outside the requirements of language. So this is my argument: as with the chaos of the Alcohol Advice Centre, I don't think it stretches common-sense to say that the collision crafts and depicts the non-coherences that produced it, the ramifications of a messy organisational and technical hinterland. Or refracts this non-coherence. Or condenses or articulates it. Terribly, in the bodies, the injuries, and the wreckage. Pain, let us allow, is indeed a witness. Elaine Scarry reminds us that torture tells of what produced it, somehow or other, however inarticulate it may be. However much, it is precisely about taking words away (Scarry 1985).

To be sure, a terrible accident is not a material form for allegory that anyone would want to foster. There are allegories and allegories, and this is too dreadful to play with. But what is at stake is not the creation of horror. Rather it is about how to think about it and what to do with it when it happens. To read it as enacted by a single set of causal circumstances. That is one possibility – an option followed in the inquiry. To acknowledge a set of non-coherent realities that escape a single narrative – that is an alternative. The making of pain, broken lives, lost partners, parents and children, these are the kinds of realities we apprehend if we read the wreckage more directly. If we acknowledge and apprehend these realities materially, corporeally, and emotionally. The argument, then, is that the coherences of textuality make powerful realities, but they also lose something: the non-coherent, the non-textual. Realities enacted in other ways. And if we simply stick with the textual then we stop ourselves from 'reading', from knowing, from appreciating, those realities. Those may be cruel realities, but a politics that does not apprehend and make them is also the enactment of its own exquisite form of cruelty.[95]

Gathering

To summarise. In practice what is present is always treated allegorically. It is read to see what it can tell us indirectly about absence. Representations and statements are no exception. Signs that tell directly about what they describe did not do this when they started life: they too were read as symptoms, indirect messages in need of interpretation. If representation is particular it is because it denies its origins in allegory, Othering the mediations that have produced its apparent transparency.

So allegory is denied but it is ubiquitous. Even more important, it is also *generative*. It messes with the boundaries between manifest absence, visible realities that can be acknowledged, and Otherness, those realities that are also being enacted but rendered invisible. It extends visibility – or it crafts and plays with different versions of visibility. By the same token it extends realities – or it crafts and plays with different and alternative versions of reality. So it is a mode of discovery – perhaps it is *the* mode of discovery. It is a set of tools for making and knowing new realities.

But there is something else too. Allegory is tolerant of ambiguity and ambivalence. Let me put the point more strongly. Allegory is *made* in ambiguity and ambivalence. To work in allegory is to see and to make several realities at once. It is to see and make several different realities in the same presence. A statement about the world is also (for instance) a statement about the motives of the person making the statement. Their social interests. Their psychiatric state. Their lack of breeding. Or their ignorance. Allegory is necessarily, then, about piling different realities up on top of one another. It is about the *apprehension of non-coherent multiplicity*. It is about split vision. Or ways of knowing in tension.[96]

Do we want to apprehend and enact non-coherent multiplicities? Euro-American metaphysics, in so far as they are carried in natural and social science, usually say 'no'. Or, to be more precise, they propose a division of labour between science and art. Or between external realities and personal experiences. Poetry or painting or novels may escape the requirements for coherence and consistency because their 'out-there', the absence that they enact, is not taken to be 'real'. It is not 'really out-there' – and in the imagination non-coherence is allowed as a possibility. So individuals are authorised to dream without any requirement of consistency. But realities are more serious. They demand singularity, and singularity demands experts, a single point of view. Non-coherent realities disappear into art, or the realm of the personal.

Non-coherence is not necessarily a good. Witness the disorganisation of the regime for treating alcoholics, or the lack of co-ordination in the railway system. But neither is coherence necessarily a good. Witness the bush-pump, or the cervical screening programme. This means that the question, 'do we want to apprehend and enact non-coherent realities?' should not seek a single response. Instead it needs to be answered case by case. However, the problem with Euro-American metaphysics is its lack of symmetry. It simply assumes that coherence is a good, and tries to enact it into being. It makes no space for the acknowledgement of non-coherence. It makes no space for allegory that knows itself as allegory. And it also *enacts coherence in a very particular way*.

It is like this. If allegory is tolerant of non-coherence, then we might also ask, what *is* it, this 'non-coherence'? What *is* 'inconsistency'? Once again we are in territory that has been mapped for us by Annemarie Mol, the territory of difference. Are the different enactments of atherosclerosis in the hospital that she visited 'consistent'? Well, the question has to be answered not simply by looking to see whether they fit a smooth and singular narrative of the kind offered by the textbook. It also has to be answered by looking to see what is happening here and there in practice. Thus the hospital moves along in its daily practice, and different enactments of the condition rub along together.

They rub along together in a variety of different ways. Some of these preserve a single coherent narrative in one way or another. Sometimes – for instance when contra-indications turn up in the case conference – they subvert it. But other times – for instance mutual exclusion or the fact of being located in different places – they do not. Often different realities are simply held apart. The 'consistency' of the object is not being tested at all. But overall it nonetheless coheres.

This is why I have preferred to talk of coherence and non-coherence rather than of consistency or inconsistency. The word 'consistency' bears a heavy weight because it draws on the particular demands of logic or discourse. It is intolerant of difference or multiplicity. These are easily turned into signs of inconsistency or incompatibility. Sometimes, no doubt, things are indeed incompatible. Two trains cannot try to occupy the same volume of Euclidean space without disastrous consequences: this is a dramatic enactment of incompatibility. But coherence – or non-coherence – is more permissive. Indeed more than that. Non-coherence may be what keeps the system held together. Singleton's argument about the role of ambivalence in the cervical screening programme shows this. But the argument can also be applied to the railways.

A final story. Earlier in this chapter I mentioned the Driver Reminder Appliance. I said that it may have been a contributory factor in the accident. I also mentioned that it was a safety device. Intended to stop drivers absent-mindedly starting after a halt, on stopping the driver would press a button to illuminate a small signal in the cab and disconnect the lever for applying power. To apply power again the driver would actively have to turn off the device.

The rationale is self-evident. It 'reminds' the driver not to apply power without thinking about it. However, at the inquiry it became clear that the device was being used by drivers in trains that were moving to remind themselves whether the previous signal they'd passed was green. For instance drivers would press the appliance if they passed an amber signal, in order to remind themselves that the next signal was likely to be red. This sounds like a sensible safety precaution. Indeed, most of the time it was. There is no evidence for this one way or the other, but it is quite possible that it may have prevented serious accidents on previous occasions. However there are circumstances in which it reduced safety. Imagine a driver passing through an amber light and *forgetting* to set the appliance. Imagine that the train arrives at the next signal which is red. Then imagine that the driver does not see that it is red. What does the appliance say? Since it has not been set, the message is that the previous signal was green, so the signal just passed cannot have been red.

It is possible, even quite likely, that this is what happened at Ladbroke Grove. Certainly the argument was made in the inquiry. The rule book says that the appliance should not be used when trains are moving. But lots of drivers used it in the way I've just described. And they were doing so not in order to cut corners, but to *increase safety*. This is a local adaptation or variation or non-coherence that isn't easy to defend given its possible importance in the Ladbroke Grove collision. But most of the time it worked very well: it may, as I have noted above, have prevented accidents on other occasions. But here is the oddity. In the single reality created in an inquiry, ambiguities and ambivalences – including local changes to the rules – are treated as one of the causes of the accident. The possibility – indeed the probability – that often enough under other circumstances they are crucial in securing workable coherence disappears. Singularities are not only sought, but they are normatively enacted. A good reality is one that is centrally co-ordinated. Non-coherent realities such as bush-pumps, health screening programmes or decisions to cancel aircraft are poorly appreciated – or they look like mistakes. This is why we also need allegorical methods.

But what to call these methods? How to think about the elements that they bundle together? In the account offered above I have mobilised a number of metaphors. For instance non-coherence. But I have also been uneasy. Sometimes I have wanted to say that the local adaptations and 'non-coherences' produce their own form of coherence. At this point we cannot avoid a debate in the politics of terminology. A term such as 'non-coherence', though (and deliberately) not the antonym of coherence, is nonetheless caught in the connotations of a standard binarism, the distinction between coherence and incoherence. Obviously that binarism values coherence. Incoherence is likely to be a bad. Consistency and inconsistency enact an even more insistent and asymmetrical binarism. This is why I have mostly avoided using this pair of terms. So the politics is complicated. Assuming the need for a more generous version of method, one can imagine at least three ways of handling these politics. First, one could insist that coherence is a good, but conventional methods are much too restricted in the way they imagine it – which is why we need a self-conscious commitment to allegory. Second, one could equally well insist that coherence is not a good – and that it is another framing feature of Euro-American method that would be better undermined. That methods, in other words, should be allowed to make non-coherences alongside coherences. This would be a more radical position, and indeed one that is tempting. Third, one could try to avoid the issue by finding a way of talking that does not leave hostages to fortune one way or the other. And this is why I have called this section 'gathering'.

To talk of *gathering* is to mobilise a metaphor that is similar in some ways to the bundling in the broader definition of method assemblage. To gather is to bring 'to-gether'. To relate. To pick (as with a bunch of flowers). To meet together. To flow together. To have, as the Quakers put it, a 'gathered' meeting for worship.[97] To build up or add to (as with a gathering storm, or gathering darkness, or a gathering boil). Gathering, then, has its own connotations. But it tells us nothing of consistency or inconsistency. And nothing of coherence, incoherence or non-coherence. Here, then, it is symmetrical.

INTERLUDE:
Notes on symmetry

The sociology of science existed as a discipline before 1962, the date of the first edition of Kuhn's *Structure of Scientific Revolutions*. As I noted earlier, it was invented by Robert K. Merton, was empiricist and positivist in its understanding of the proper relations between science and reality, and in particular articulated the idea that science needed protection from distorting social and political influences. But after the appearance of Kuhn's book the sociology of science turned itself into a *sociology of scientific knowledge*. The emphasis was on *knowledge*, and the discipline became a version of the sociology of knowledge. The guiding idea was that scientific knowledge is a form of culture that is shaped by social and economic interests – and then that this process of shaping is not necessarily problematic. The argument is most clearly articulated in the writing of Barry Barnes (1977) and David Bloor (1976). Barnes and Bloor note that paradigms are puzzle-solving tools for handling reality, and that such tools can be understood as cultures. Passed on within a scientific community, in principle they are like any other set of beliefs and tools which the community in question uses to make sense of and to live in the world. And if this is the case, then they can also be analysed as cultural forms.

But how are cultures shaped? Kuhn offers an 'internal' explanation. He looks at how scientists set themselves paradigm-defined puzzles, and argues that paradigmatic culture grows as a function of successful puzzle solving. He is not very interested in 'external' social factors – these are not his primary concern. But, say Bloor and Barnes, external social interests may also be important. Indeed, very often they are. Their position, then, is that scientific tools are shaped both by natural reality and social (including professional puzzle-solving) interests.[98]

But how to study the social shaping of scientific culture? What methodological approach is appropriate? It is, say these authors, essential to avoid what is sometimes called 'Whig' history. A Whig history is one that explains the past in terms of its contribution to the present. In the context of science, such an explanation would use present-day scientific thinking as a guide to explaining past scientific progress. In practice it thus treats past scientific knowledge asymmetrically. Knowledge that fits with what scientists now believe is in no need of further explanation because it is true. Knowledge that does not fit with current ideas, and is therefore now taken to be wrong, does, however, need to be explained. It needs to be explained because some explanation is needed for the fact that the scientists in the past failed to understand that it was true. This, then, is an *asymmetrical* explanation. It is asymmetrical because true knowledge and false knowledge are explained in different ways.

This won't do, say Bloor and Barnes, because if science is a form of culture and we want to understand what is going on, then our judgements of what is true and what is false are irrelevant. In the first instance, at least, we need to judge the culture in its own terms. In general people act rationally – this is the assumption – given their circumstances and their cultural resources. This means that if we want to explain how people act and what they believe, then we need to understand how *they* picture the world. Our own judgements about reality are irrelevant.[99] So what is needed, methodologically, is *impartiality* (with respect to truth and falsity), and *symmetry*. Thus Bloor writes of a sociology of scientific knowledge that:

> It would be symmetrical in its style of explanation. The same types of cause would explain, say, true and false beliefs.
>
> (Bloor 1976, 5)

All beliefs, true and false, are shaped, says Bloor, by the natural world on the one hand, and the operation of social and psychological factors on the other. We should be explaining them in the same way.

Barnes, Bloor, and the other sociologists of scientific knowledge thus offer a theory of scientific entanglement close to that proposed by Kuhn, but they add in 'external' social factors. However, it is the notion of symmetry that is most important here. Thus the idea that symmetry is a methodological good has been extended by other writers. In particular, sociologist of science and technology Michel Callon has turned it into an ontological argument (Callon 1986, 200). In an approach that is close to that of Latour and Woolgar, he says that investigators should offer the same kinds of explanations for events in the natural and in the social worlds. The argument is that since both nature and culture are being produced together and in the same process, it is unsatisfactorily asymmetrical to assume that (say) nature has a particular and distinctive form, and therefore needs to be explained in terms that are different to those of the social. Rather, says Callon, we should follow a principle of 'free association':

> Instead of imposing a pre-established grid of analysis upon . . . [the entities and their relationships mobilised by actors in discussion], the observer follows the actors in order to identify the manner in which these define and associate the different elements by which they build and explain their world, whether it be social or natural.
>
> (Callon 1986, 201)

This, then, can be seen as the extension of methodological symmetry into ontology, into what there is. What there is and how it is divided up should not be assumed beforehand. Instead it arises in the course of interactions between different actors. But note also that for Callon what counts as an actor can only be determined in the course of interactions. Actors are entities, human or otherwise, that happen to act. They are not given, but they emerge in relations.

So if Bloor and Barnes recommend epistemological symmetry, then Callon is pressing the case for ontological symmetry. And what I want to do is to link the two suggestions together. This is because to imagine method assemblage in a generous and permissive way we require both. For as we have seen, to talk of method assemblage is to say nothing about the character of absence, the condensations of presence, or the mediations that produce these. It is deliberately permissive. So this is a version of ontological symmetry. The principles of symmetry and free association are being extended *to the character of method itself.* For if we want to understand how understanding is, or may be, achieved, we should not distinguish in the first instance between good and bad methods assemblages. In particular, we should include methods assemblages that:

- enact absences as independent, prior, singular and definite, and those that do not;
- craft presences or condensates as representations, allegories, objects and events.

Asymmetry with respect to methods assemblages limits the realities that can be known, and forms by which we can know them. My argument is that this is epistemologically, ontologically and politically inappropriate. Judgements about method need, instead, to be made in ways that are specific and local.

6 Non-conventional forms

Introduction

> We are not good at thinking movement. Our institutional skills favour the fixed and static, the separate and self-contained. Taxonomies, hierarchies, systems and structure represent the instinctive vocabulary of institutionalised thought in its subordinating of movement and transformation. The philosopher Whitehead (1925) called this the principle of simple location in which clear-cut, definite things occupy clear-cut, definite places in space and time. There is movement – of a kind: the simple movement of definite things from one definite place to another. But it's a form of movement which denies the restlessness of transformation, deformation and reformation. Simple location reconstitutes a world of finished subjects and objects from the flux and flow of unfinished, heteromorphic 'organisms'.
>
> (Cooper 1998, 108)

The scheme I am proposing is permissive, and in the first instance the approach is ontologically and epistemologically symmetrical. Method assemblage, I am suggesting, may gather in any form. The absent out-therenesses enacted may be, but do not need to be, definite and singular. The condensed in-herenesses crafted into presence may take, but do not need to take, the form of statements or texts. But what *are* methods assemblages, over and above a series of mediations that produce presence, manifest absence and Otherness? Is there anything more that we can say about them? In this permissive turn to the ontological and its politics, what is being implied about absence? These are the questions that I explore in this chapter.

I work once again through empirical materials – in this case three small case studies. These are, first, the development of a project management system; second, a laboratory ethnography; and third, a Quaker meeting for worship. Each is a form of method assemblage. I explore these cases, and the more general metaphysical and methodological questions, by addressing a particular and very matter-of-fact problem experienced by many natural and social scientists in the course of their research. This is the paradoxical experience that, on the one hand, and at least some of the time, reality seems to be overwhelming and quite dazzling. And then, on the other hand, the contrary experience that there is not much of interest going on: that somehow or other, at some stages in research, the world has gone silent. These contrary but related experiences are, I suggest, a key to the character of the method assemblage and the metaphysics in which it is situated.

Daresbury SERC Laboratory

In 1990 I spent an ethnographic year watching the work of managers, scientists and engineers at Daresbury SERC Laboratory. This was a major, largely publicly funded, scientific facility near Warrington in Cheshire with more than 600 employees.[100] At the time it ran several research facilities (the largest was a synchrotron radiation source) which were available, usually on a grant-awarded basis, to visiting scientists from British and overseas universities and, increasingly, to large science-based companies such as ICI. Users would come and set up experiments which made use of experimental 'beam time' at one or other of the many experimental stations.

During the year of my study I watched researchers conducting experiments, sat with the technicians responsible for running the facilities on a day-to-day basis, and interviewed a wide range of staff. I also attended most of the major management meetings. I will return to the ethnography itself below, but first I want to discuss the process by which managers built a project planning system. In one paper about this I recreated an ethnographic present about a particular incident to do with project planning in the following terms:

> Andrew is sitting at his desk. He's about to call an emergency meeting of the management board. He's bothered because the 'second Wiggler project,' the so-called 'flagship project' for the laboratory, is starting to fall seriously behind schedule. But what is there to see of this second Wiggler project? Does it look as if it is behind schedule? The answer is, no it doesn't. Not really. Not in any way that you or I could see. For as he sits fretting in his office, it is nothing more than a hole in the ground, and a bunch of construction workers pouring concrete. There's no particular sign that anything is wrong. It's a mess alright, but only a mess in the way that all construction sites are a mess: hard hats, hard shoes and mud everywhere.
> (Law 2002b, 27)

Andrew is the name I gave to the then head of the laboratory. And what interested me about the story was the way in which Andrew was able to see something that the rest of us could not: here a very serious but otherwise invisible delay to this important 'second Wiggler project' – a project with a budget of several million pounds that was vital to the future of research based on synchrotron radiation research, and indeed to the laboratory itself.

So how did he achieve this apparent ability to see the invisible? The answer is that he enjoyed the benefits of a spreadsheet. This arrayed figures that represented the amount of effort going into the project, which in turn allowed him to make a quick comparison between the actual effort and the amount of effort that was *supposed* to be going into the project. And the particular morning which I report above had revealed a worrying truth. This was that according to the project plan eighteen man-years of effort (such was the laboratory vernacular) should have been devoted to the project at this point in its lifetime, but in reality only eleven man-years had actually been put into it. Though this wasn't visible to the casual observer, the discrepancy told him that the project was developing at only two-thirds of the intended rate. Indeed it revealed that all the contingency time built into the original project plan had already been used up. In short, it was months behind schedule, and this fact was likely to become dramatically visible eighteen months or two years on, when the second Wiggler was supposed to be completed and ready for users.

So how did Andrew know all this? I have mentioned the spreadsheet. As is obvious, a spreadsheet is a device for arraying, juxtaposing, relating, creating, and (at least very often) simplifying numbers. If Andrew lay at the heart of a method assemblage, then the spreadsheet was producing a set of statements or inscriptions that (were said to) correspond to a reality. But, as is also obvious, the spreadsheet was simply a small part of a larger inscription device, a set of mediating practices crafting not only a set of figures but a manifest reality, a hinterland out-there. So what of that absent hinterland?

A part of this involved what the laboratory called the 'manpower booking system'. The basic idea was simple. Each employee was supposed to fill in a form every month to describe how he or she had spent his or her working time. The form was a compromise between precision and robustness. Thus the grid of possibilities available to employees was relatively coarse: time was broken into half-day units, rather than quarter days or hours, and the employees were given only a few codes to enter in those half-day slots. These codes mainly referred to a series of projects (the second Wiggler was one) plus a few residual categories, including general laboratory administration and management.

Some employees resented the need to fill in these newly invented forms and invented spurious uses of their time or refused to file them altogether. However, most filled them in, albeit with a fair amount of grumbling. Life, they pointed out, does not naturally fall into half-day segments. Many mornings are fragmented. Some tasks are relevant to more than one project. And what about all the unclassifiable bits and pieces that seem to take up so much time in a working day? Nevertheless, in the end most employees returned their completed forms through their project leaders to the accounting department where they were checked for glaring anomalies. Then the laundered results were entered into a spreadsheet. And this was the point at which they became arithmetically tractable in relation to projects and project effort, the moment when senior managers such as Andrew could draw out and explore manpower figures.

Dazzling and simplifying

This is method assemblage in a conventional representational form. It bundles relations together to produce an organisational reality out-there on the one hand and a set of traces in-here on the other. Most of the work of assembling them has been Othered by the time the figures reach Andrew's desk, and the inscriptions are taken to correspond to organisational features that are enacted as independent, prior, definite and singular. So though the managers had a pragmatic attitude to the manpower booking system (they knew that it was rough and ready, a tool for doing a practical job rather than a perfectly tuned instrument) they also took it for granted that for practical purposes reality was fairly simple. At least much of the time they assumed that people work on this project or that, and this was something that could be measured. It was the broad shape of effort that they were interested in, and not precision. Complexities and details would simply get in the way.

As indeed, at other moments in the laboratory, they did. For instance, there were meetings where the managers arrived with print-outs that had clearly been spewed out by a line printer. These fanfold print-outs were huge – sometimes comprising an inch-thick array of A3-sized pages. It turned out that they detailed all the costs (invoices, services, contracts, salary components) charged to each project. So what did the managers and project leaders make of this mass of paperwork? The answer is that they regularly complained that they were being drowned in detail. They said they were supposed to be *controlling* their projects, which meant that they were concerned with overall spend, with setting priorities, and with overall manpower or purchasing. But the print-outs gave detailed information about all sorts of irrelevant specificities. One manager complained it had taken him a whole evening to plough through his print-out, and he still didn't have an overall grasp of the shape of the spending on his projects. He commented that he simply didn't need to know the details that it enumerated: that indeed they got in the way. Yet there it was: three boxes of number eight countersunk steel woodscrews requisitioned from the stores on 19 September. Very interesting but entirely irrelevant to the bigger picture.

What is the importance of this story? One answer is that it shows that as a part of making and condensing realities, method assemblages necessarily *craft complexities and simplifications*. Andrew's manpower spreadsheet and the big print-outs make both in the course of enacting project realities. Andrew's spreadsheet is simpler but despite their disabling complexities the big print-outs are *also* simplifications. They work by ignoring most of the events that make up the laboratory, and attending to and building upon very specific patterns of events. The general lesson is that to enact out-thereneses is to make *silences and non-realities as well as signals and realities*. This double movement – realities made and realities unmade – is constitutive of method assemblage.

But here are some more data to think with, my second case.

At the beginning of my year in Daresbury I found that I was constantly being dazzled. There was too much going on. Meetings, activities, experiments, disasters, triumphs, comings, goings, arguments, friendships, documents, policies, programmes, aspirations, promotions, conferences, memos, cups of coffee – all of these and much more were included in the daily round of laboratory work. And since the site was large, and many of the activities of the laboratory ran day and night, they were also amply distributed across time and space. The effect was overwhelming – a bit like the experience of the managers with their inch-thick print-outs. Sometimes, especially in the early days of the ethnography, I found that I needed to retire to my car to eat my sandwich by myself at lunch time, or to use the library to make some peace.

At the time I tended to think that this was my own particular problem: that I wasn't coping properly with the incessant demands of ethnography. I wondered if a better ethnographer would have been on top of all the detail and better able to keep track of the ethnographic equivalent of boxes of wood screws. However, I now think that something much more interesting and important was also going on. It was that in the ethnographic method assemblage *the practices that I needed to make certain silences and unrealities were not in place*. I was being overwhelmed by the presence of too many inscriptions or traces in-here, and the manifestation of too many realities out-there. Too many realities – and representations of realities – were being enacted. In short, as with the print-outs, the balance between the manifestation of entities, the real, on the one hand and the enactment of the non-real, of silence, of Otherness on the other, was wrong. Allegory is about enacting, and knowing multiple realities. But as I suggested in the last chapter, allegory is also about the movement between realities. In particular, it is about holding them together. To misquote T. S. Eliot, there was too much reality to bear.

If this diagnosis is right then what I needed was a *better tuned* and more discriminating method assemblage. I needed to make a version of coherence by re-working the boundary between manifest realities and Otherness. Some of the tools that I needed were in place. For instance, instead of tape-recording I made notes. This meant that much was being routinely Othered, including gestures, tones of voice, and most of the physical surroundings. At the same time my many pages of notes were detailed, and included numerous near-quotes. So the ethnographic method was assembling a condensate of traces. And certain repetitive patterns – like words, sentences, meetings, topics and agendas – were condensing themselves into those notes and enacting a corresponding version of laboratory out-thereness. However, as with the financial print-outs, the notes were condensing too much and making too much reality. But then, as time passed, things started to change. The ethnographic dazzle started to diminish in part because I began to note different kinds of patterns in laboratory reality. But what does it mean, to talk about 'patterns'?

As we have seen, Kuhn tells us that to be a scientist is to recognise similarity between instances even though no two instances are ever the same.[101] Scientists (and other people too) creatively detect and select appropriate similarities between instances whilst ignoring others. Latour and Woolgar say something similar. Inscription devices make traces which sometimes map on to one another to produce a sustainable set of similarities. Again, the metaphor is about the need to find or make a *pattern* against an endless background of noise. Indeed, both participants and observers of contemporary scientific inquiry frequently talk of the making or finding of patterns. It is, for example, near to impossible to detect the patterns made by solar neutrinos against the background of other noises (almost all solar neutrinos simply pass, undetected, through the earth which is almost invisible to them – and vice versa).[102] In general it is exceedingly difficult to make and detect patterns which correspond to theory about elementary particles (which is why funding councils have to pay hundreds of millions of dollars to produce the appropriate detectable patterns against an overwhelming background of inappropriate similarities).[103] There is simply too much sub-atomic dazzle. Or, again, it is incredibly difficult to detect the gravity waves that are produced, in many versions of cosmological theory, by catastrophic events early in the history of the universe. Here is sociologist of scientific knowledge Harry Collins, writing about the detection of such gravity waves:

> the predominant approach to the detection of the radiation has been to try to integrate the energy of the radiation in a device that will vibrate naturally at the same frequency as that of the putative wave. . . . [In one such case] the integrating ring was a large bar (several tons) of aluminium alloy which would 'ring' at a characteristic frequency. . . . Vibration in the bar would be detected by piezo-electric strain gauges glued onto it, their output amplified and recorded.
>
> (Collins 1981b, 35)

But Collins shows how this is only the beginning, since the bar can be expected to vibrate as a result of disturbances that have nothing to do with gravity waves:

> . . . the bar must be insulated from all other known potential disturbances. Electrical, magnetic, thermal, acoustic and seismic disturbances must be guarded against. . . . [The experimenter] attempted to do this by suspending the bar in a vacuum chamber on a thin wire. The suspension was insulated from the ground by a series of lead and rubber sheets.
>
> (Collins 1981b, 35–36)

And this is still only part of the story. For instance, since it was not possible to cool the bar to absolute zero, the strain gauges picked up endless signals that had nothing to do with gravity waves but instead reflected random thermal movements of atoms. A signal representing a gravity wave would thus be found among signals caused by atomic thermal movement. Collins adds:

> A gravity wave would be represented by a particularly high peak . . . , and a decision has to be made as to a threshold above which a peak counts as a gravity wave rather than noise. However high the threshold that is

chosen, it must be expected that occasionally a peak due entirely to noise would rise above this threshold.

(Collins 1981b, 36)

So the detection of gravity waves was also a matter of statistical manipulation and judgement. Experimentalists needed to show that high peaks occurred more frequently than would be expected as a result of random thermal noise.

Collins's description (and many other comparable studies) show that making and detecting 'the right' similarities and differences is difficult, complex, and involves going to extraordinary lengths to delete 'the wrong' similarities and differences. This is because there are just too many possible similarities and differences out there. What we think of as, or come to call, 'noise' is – all those 'wrong' similarities and differences. The implication is that *realities grow out of distinctions between 'right' and 'wrong' patterns of similarity and difference*. It is this that enacts the distinction between real and unreal, and makes signal and silence. The implication is that silence and non-realities are also artful effects. They are the first steps towards avoiding dazzle and making realities.[104] Specific out-thereness depends both on the Othering creation of silence and on very selectively attending to, amplifying, and so manifesting, possible patterns.

How does this apply to the Daresbury ethnography? Let me pick up the question empirically. Early in the study I asked participants how the laboratory had changed over the previous decade. At first overwhelmed by a lot of detail, as I listened to their responses, I became impressed by what I came to see as two different versions or styles of that story. The first was gradualist. In this the history of the laboratory was described as an evolution, an accretion, a process that developed progressively, step by step, to reach the point where the laboratory had achieved its current level of success. This was in stark contrast with a second heroic style of narrative which stressed discontinuities. This said that the laboratory had been in mess, rudderless, ineffective and drifting. Then, *in extremis*, it had been saved by the arrival of a new and entrepreneurial management team that had quickly and decisively taken the problems in hand and 'turned the laboratory round'.

How to think about this? The answer is that the interviews contained limitless possible patterns of similarity and difference. Limitless possible realities. This was the dazzle. Amongst these, however, were the two narrative styles, and these fairly quickly became the 'right' pattern, the one to attend to, to discover, and to amplify. How did this happen? The answer is partly empirical. The relevant patterns were, of course, discoverable in the materials I gathered. In particular, however, they were discoverable in a rather stark stylistic distinction between two of my initial interviews. As it happened one of these was dramatically heroic while the other took a gradualist and undramatic form. I initially found it difficult to reconcile the two. This was a puzzle: how to bundle them together and make a story?

A part of the answer is that these similarities and differences resonated with another quite different set of possible similarities and differences that are rehearsed and amplified in one of the possible theoretical hinterlands. This is a long-standing literature in the sociology of knowledge which insists that there are dramatically different and socially shaped understandings of history. These understandings are – yes – heroic, philosophically *romantic, and discontinuous* on the one hand, and *evolutionary, rationalist and incrementalist* on the other.[105]

The result was that data and theory interacted together in a way that resonated and amplified one another to produce pattern and repetition. These two interview narratives could be seen as signs or instances of the two great narratives of history. And with this pattern resonating it became progressively easier to find additional ethnographic moments that might be understood as further repetitions of the same pattern. As a result my field notes suddenly started to produce signals. What had been dazzle, an overwhelming out-thereness, was converted into signal on the one hand and silence (which did not resonate with the relevant pattern) on the other. And the same logic applied to new notes and field observations. Bits and pieces in those observations became instances of repeatable patterns and signs of the dual heroic/incremental discursive reality of the laboratory and its ordering. At the same time other bits and pieces became less significant. The signal grew against a growing background of silence. Indeed, in due course I found it difficult to attend to forms of talk which did not fit this basic pattern of repetition.

Modes of ordering

What is the character of organisation? What is the structure of a large laboratory? What is it to manage an organisation? These were the questions that I was trying to answer as I spent time at Daresbury.[106]

To answer them I needed to cut through the dazzle and Other almost every possibility. This started in the process that I have just described. Fairly quickly, patterns started to emerge from and resonate with the data assembling in my notebooks. The process of Othering, of deletion – which was also, and in the same moment, the process of pattern-making – was under way. A more tractable reality was being enacted.

But there were two patterns. One, as I have noted, was classical, incremental, evolutionary, and decidedly undramatic. Events in the laboratory had unfolded bit by bit. Certainly there had been problems, but there are always problems in any organisation. And the task, as always, was to solve those problems, treat them as puzzles to be resolved. And indeed they had been solved. The laboratory had moved on. The second pattern was heroic, romantic, and discontinuous. It was about qualitative change. Again as I have noted, this conceived of the history of the laboratory as a dramatic 'before' and 'after'. Before, it was beset with difficulties, on the verge of catastrophic failure. Indeed it was about to be closed. But then the new management team had been drafted in. The result was a dramatic discontinuity. The problems had been addressed. Things had been put right. A new urgency and dynamism was injected into the organisation. The laboratory was put back on track. Its future was assured. It had been saved.

So there were two histories to choose between or combine: the classical and the romantic. How to do this? How to reduce the dazzle still further? After a while I concluded that neither was correct. Or better, I concluded that they were simplifying stories and that the history of the laboratory was more complex than either. At the same time they were not stupid. Perhaps, then, both were partially right. Or, to put it differently, perhaps the history of the laboratory could be understood as *the enactment of both*. This suspicion was confirmed when I started to find that these two patterns – I came to call them 'administration' and 'enterprise' – also repeated and resonated in quite different contexts. They were two quite different styles for decision making that seemed to co-exist, often within the same person. Sometimes they were in conflict. According to administration, enterprise often broke the rules. It was too attached to the main chance. It was a 'cowboy' logic. Conversely, from an entrepreneurial point of view administration frequently looked like 'civil service' pen-pushing, more concerned with due process and form-filling than responding to the challenges of the real world. At the same time, often enough they depended on one another (enterprise needed the legalities of due process, while administration depended on the more responsive approach of enterprise).

In due course I came to the view that the organisation of the laboratory was not any single thing. It wasn't simply entrepreneurial. Neither was it simply administrative. Both of these – I came to call them 'modes of ordering' – were being enacted in and enacting the structure of the laboratory. Indeed, in due course I further concluded that there were other modes of ordering too. For instance, there was a pattern of charisma at work with its own specific organisational logic. And also that there was a good deal of Kuhnian-style puzzle solving too. And organisation (a verb rather than a noun) was the enactment of all of these and their different interactions (and a lot more besides). Organisation, then, was multiple. It was multiple patterning, multiple versions of repetition, and multiple modes of Othering.

By now it is clear that everything said by Mol about multiplicity also applies to organisation. But so too do the arguments about allegory. For the managers can be understood as consummate allegorists. They lived, enacted, depicted, in short, they gathered – a series of different and (non-?)coherent realities. Perhaps organisation itself *is* allegory. It is gathering. Perhaps it is the creation, recognition and tolerance of different patterns altogether. A process of holding things together that are not strongly consistent. Perhaps, then, good organisational studies are also studies in allegory that depict and manifest realities that achieve allegorical gathering rather than a single-version discursive consistency. Fractionality.

Quaker meeting

My argument is thus that the practices of *method assemblage craft out-thereness by condensing particular patterns and repetitions whilst ignoring others*: that they manifest realities/signals on the one hand, and generate non-realities/silences and Otherness on the other. Unless they do this representationally or allegorically then they fail. They are overwhelmed by dazzle. Indeed Collins describes just such a failure for the case of gravity waves. Science/silence: to make realities is to unmake possible realities, endless numbers of them. But which?

In this book I have talked primarily of method assemblage in natural science and social science. In the present context their similarities are more important than their differences. But as we have also seen, other practices – health care or train collisions – also craft realities, depictions of those realities, and Othernesses. They also make and select between 'real similarities' and 'unreal silences'. Consider, for instance, the following, which describes events in a room:

It is modestly furnished. Modestly decorated. The people are variously dressed, many of them quite informally. They're sitting on upright chairs, in a rough circle. There's a small table in the middle, with a bunch of flowers. And a few books. But otherwise, there is nothing. No furniture. No movement. No talk. For it is the silence that you're going to notice most of all.

Some of the people have their eyes closed. A few are staring, in an unfocused way, at the flowers, or beyond the flowers to the people on the other side of the room. Or out of the window where, if you look, you can see distant rooftops and clouds. And as you listen in the silence, the loudest noise is the call of children from a nearby garden. Or the sound of a car passing in the road.

How to convey the character of that silence? It isn't heavy and preoccupied, like the desperate hush of an exam room. Nor is it disciplinary and repressive, like the pressure that expands to fill the space of the parade-ground where you hardly dare breathe. It isn't the silence of a graveyard with its imagined echoes and distant memories. Nor is it the silence you hear when you lie in the breeze on your back in the sun on the turf of the chalk downs. None of these, though perhaps the last comes closest to it. Instead it is, as they say, a 'centred' silence.

(Law and Mol 1998, 20)

This describes a Quaker meeting for worship. The Quakers are a small group within the Christian Protestant tradition that trace their origins back to the 1650s. They have no ministers, or priesthood, and no permanent appointments. Instead they govern themselves – or, more properly, they allow the Holy Spirit to govern them. Quakerism is a theocracy, though it is easily mistaken for a democracy. Anyone at all can attend Quaker meetings, and members have few or any special duties or privileges. Those who seek membership are asked to attend to the Christian tradition and the questions that it raises, but they do not necessarily need to believe the specifics of the Christian tradition. Indeed, membership does not require belief in anything in particular at all. A concern or a sensibility to the spiritual is more or less all that is implied. So it is possible to be a Quaker and, say, a Methodist, a Buddhist or a Pagan. The divine reveals itself in many forms and modalities, say the Quakers. There is no monopoly, no correct way.

In the Quaker world divinity is everywhere. Immanent and transcendent, it is in the lives of people, the world of work, and in commerce, nature and personal friendships. This is not to say that everything in the world is good. Indeed, there is much that is bad, and many Quakers are committed to politically radical versions of political, economic or charitable work as a witness to God's work. But it is the ubiquity of the spiritual that explains why most Quakers will not, for instance, swear on the bible when they give testimony in court. (To swear on the bible would be to imply that there is something special about what follows but *everything* is special and carries the divine.) It is why many Quakers are pacifists. (There is that of God in every person, so it is not easy to see how it could be right to kill anyone.) It also explains why the Quaker meeting house is a quiet, plain, under-furnished, undemonstrative place. In principle there is nothing special about the place of worship since divinity is everywhere.

What is the form of worship?

When I first went to a Quaker Meeting I wrestled with the questions that happened to be bothering me. This went on for several weeks. But then I learned that this wasn't what silent worship was about. For after a time somebody came and sat next to me. And at the end of the meeting she started a conversation which led me to ask how I should worship.

A: 'You let the thoughts swim by you.'

Q: 'What do you mean?'

A: 'Think of it as meditation. You are being distracted by all these thoughts. Ideas keep on popping into your head. What should I cook tonight? Who do I need to phone? You can't stop thinking these thoughts. But what you *can* do is to take them, just take them, like

a fish, and throw them back into the river. Stop thinking. Not by
forcing yourself to stop thinking. That will never work. But by
embracing the thoughts and simply letting them go.'

(Law and Mol 1998, 23)

To worship in this way is to find ways not to be distracted by all the noise,
literal or metaphorical. Many Quaker writings reflect on this:

> Sometimes . . . the prayer following meditation leads to an inner silence,
> a stillness in the depths, which is the peace of God, passing all under-
> standing. It cannot be commanded at will for it is the gift of God, a
> blessing which he gives only to those who can cease from anxious striving
> and desiring. Some of us, alas, have known it only on a few occasions, but
> these are our richest memories.
>
> (London Yearly Meeting of the Religious
> Society of Friends 1960, 251)[107]

This is why the unprogrammed Quaker meeting for worship is mostly silent,
with people gathered, sitting, meditating, reflecting, perhaps praying silently.
They are waiting to hear and to be moved by the Holy Spirit. Sometimes
the silence may last for a whole hour. More often it is broken as a member
of the meeting rises to her feet because she feels the need to offer 'spoken
ministry':

> When one rises to speak in such a meeting one has a sense of *being used*,
> of being played upon, of being spoken through. It is as amazing an
> experience as that of being *prayed through*, when we the praying ones are
> no longer the initiators of the supplication, but seem to be transmitters,
> who second an impulse welling up from the depths of the soul. In such an
> experience the brittle bounds of our selfhood seem softened, and instead
> of saying 'I pray' or 'He prays' it becomes better to say 'Prayer is taking
> place'.
>
> (London Yearly Meeting of the Religious
> Society of Friends 1960, 249, part)

A part of the reason for the silence and the prayer is to help break down what
are for many Euro-Americans the everyday habits of selfhood – the sense of
being an individual with a distinct and separate identity, and with specific and
personal goals and plans. The object is to break down the boundaries round
the person so that he or she can be 'used' by the spiritual. It is to act in it and
for it, and reflect another reality that is not always so apparent, that of the
spiritual. For the love of God, divinity, is infinite, but it is also difficult to
detect for most of us in the everyday rush of events. The question is: how to
live it; how to know it; and how to tell it:

We may not issue from a gathered meeting with a single crisp sentence or judgement of capsuled knowledge, yet we are infinitely more certain of the dynamic, living, working Life, for we have experienced a touch of that persuading Power that disquiets us until we find our home in Him.

(London Yearly Meeting of the Religious
Society of Friends 1960, 249, part)

Resonating

If we take a symmetrical approach then Quaker worship is a method assemblage, along with a manpower booking system, ethnography, the detection of gravity waves, or the conduct of scientists in the Salk Laboratory. Natural science, medical practice, social science, the making of any form of presence or experience, these are all enactments or modes of crafting the condensations and hinterlands of presence and absence. In the Quaker meeting, like the Alcohol Advice Centre and the Ladbroke Grove collision, what is made present does not necessarily take the form of a 'single crisp sentence' or a statement. Like these, then, we are in the realms of allegory or gathering as these press up against the limits set by the demands of language. Again the message is that if we stick too rigidly to statements then we will refuse reality to many out-therenesses.

The particular realities and condensates enacted in the Quaker meeting are more or less unlike those of science and social science. But as we have seen, crucial to all method assemblage is the need to distinguish signals from noise and so to create silences. Comparison between the Quaker meeting and the gravity wave experiment is instructive, for they are similar in important respects. For both, making enough silence is tricky. Each starts, then, with the problem that all sorts of louder realities are condensing themselves as a cacophony of patterns. This means that these louder worlds need to be tuned out in order to make the right reality. Both the meeting and the experiment, then, assemble practices to detect and amplify particular patterns that would otherwise be below the threshold of detectability. Other patterns, the 'wrong' ones, drown them out and are Othered. They are intended to resonate with and then to amplify those patterns, to take what is only just there, and then (as Collins puts it) to integrate them and (re)make their reality. Note that they both receive *and* they transmit. Picking up on a faint pattern, they make it stronger. They condense and manifest a version of reality, but as they condense it they re-enact it, they re-confirm it. *Method always works not simply by detecting but also by amplifying a reality.* The absent hinterlands of the real are re-crafted – and then they are there, patterned and patterning, resonating for the next enactment of the real.

In its primitive form it is therefore useful to think of out-thereness or absence as a set of potentials. It is all the possible repetitions of similarity and difference, the patterns that have been set humming and jangling in all the other and endless enactments. This means that it is also useful to imagine it as a set of impossibly complex interferences between patterns of repetition. It is the endlessly multifaceted intersection between different similarities and differences, which may join together, include one another, ignore one another, cancel one another, contradict one another, or silence one another. Which may be made present (or not) in the form of texts, inscriptions, bodies, skills, instruments, sensibilities, architectures, ghosts, spirits and angels – and all the other materialities one could imagine. Always, what is absent is a set of potential patterns that buzzes and dazzles and dances, that is too complicated to condense, to make present. That can only be condensed and amplified in the most selective ways. Crystallised. It is, therefore, excessive, unknowable, a source of energy and possibility, a 'flux and flow of unfinished, heteromorphic "organisms"' (Cooper 1998, 108). But at the same time it is partly made in the particular forms, and it does condense in particular locations.

How to think this? The answer is that since it is excessive there is no right way to think it, but many possibilities. Philosopher Michel Serres:

> The object of philosophy, of classical science, is the crystal and, in general, the stable solid object with distinct edges. The system is closed and is in equilibrium. The second object-model has flowing edges, it is the jet of water, the bank of clouds. It is a system that oscillates within wide margins – but has its own margins.
>
> (Serres 1980, 51, my translation)

For Serres these two forms or metaphors for the real – the solid and the fluid – endlessly intersect. So the real is flux, fixity, and *also* their intersection. He argues that we need a 'third object', a way of knowing that intersection:

> I believe, I see, that the state of things is more like a scattering of islets in archipelagos in the noisy and barely-known disorder of the sea, islets whose peaks and edges, slashed and battered by the surf, are constantly subjected to transformation, wear and tear, being broken, encroached upon; with the sporadic emergence of rationalities whose links with one another are neither easy nor obvious.
>
> (Serres 1980, 23–24)

This is what I am attempting, with my own set of metaphors. Method assemblage, craft, bundle, hinterland, condensate, mediation, pattern, repetition, similarity and difference, object, gathering, allegory and representation. There are no right answers. Local and temporary fixities grow, like the islets in Serres's archipelago, out of a sea of flux, and together they condition the circumstances for making new and temporary fixities. But the metaphor of resonance is useful.

Perhaps, then, it is helpful to think of method assemblage as a radio receiver, a gong, an organ pipe, or a gravity wave detector, a set of relations for resonating with and amplifying chosen patterns which then return to the flux, for the moment rendered real. And my concern in this book is not to foreclose on the realities that might be made too soon. Either procedurally (hence my argu-ments for allegory) or substantively. There is not all that much room for j-ψ particles in most parts of the Euro-American world.[108] Or the workings of the Holy Spirit. Or the indefinite. Or the multiple. Perhaps there is not enough room for ethnographic realities either. In social science the equivalents of 'what should I cook tonight?' are more real. As, too, are the forms to be taken by the proper answers: definite, singular, and all the rest. It takes considerable methodological discipline – but also imagination – to reduce the dazzle of noise and make the kind of silence that will allow the faint signal of the neutrino, or of spiritual mystery, to be revealed, made audible, and amplified. The disciplines that are currently pressed upon us tend to make the wrong kinds of silence. They tend to remake the silences of Euro-American meta-physics. But it is time for these to be questioned. This is why method is, or should not be, limited to representation. Why it is better thought of as crafting, allegory, or gathering.

INTERLUDE:
Notes on purity and hybridity

The first version of modern science grew up in the 1660s and 1670s in Restoration London. Historians Steven Shapin and Simon Schaffer, looking at the rise of the Royal Society and Robert Boyle's experimental work at this period, note that Boyle was faced with a problem: how to ensure that his experiments were 'witnessed' in a way that could carry conviction with other natural philosophers when those philosophers were unable to travel to London to witness the experiments for themselves.[109]

This was an epistemological problem: how to produce statements about the world that would carry conviction. But it was simultaneously a social problem: how to persuade other natural philosophers that Boyle, or Boyle's experiments, or Boyle's reports of those experiments, were appropriately authoritative. How to convince sceptics that Boyle was, indeed, an authority, an author. In response to this double epistemological/social problem, Shapin and Schaffer argue that Boyle devised three interrelated and mutually embedded 'technologies'.

- The first was *material* and took the form of an elaborate air-pump in an equally elaborate laboratory. The air-pump, they say, in an analysis that is closely related to that proposed by Latour and Woolgar and the other sociologists of science discussed above, produced phenomena that were otherwise unavailable, and might be interpreted (for instance) as demonstrating the elasticity and the pressure of air. As with most of the experimental arrangements we have considered, Boyle's air-pump was unreliable and temperamental. Finding the right patterns was difficult. But the air-pump and the laboratory were not simply technical arrangements. They also helped to create a social space. Thus the laboratory became a public location in which appropriate people might 'witness' the facts about air as revealed by the working of the pump. Appropriate people? Yes, indeed, for it turns out – I'll return to this in a moment – that most kinds of people were not appropriate and could not be counted as proper scientific witnesses.
- The second technology was *literary*. In practice very few people could make the journey to London to see the air-pump in action, and witness it for themselves in person. Some kind of indirect witnessing was therefore necessary if the experimental claims emanating from Boyle's laboratory were to achieve anything other than local currency. If what Shapin calls 'virtual witnessing' was to be achieved:

The technology of virtual witnessing involves the production in a reader's mind of such an image of an experimental scene as obviates the necessity for either its direct witness or its replication.

(Shapin 1984, 491)

This, he adds, is also a technology of trust. So how was it done? The quick answer is that Boyle created a particular kind of text that included: an image of the pump (a 'representation of reality'); a verbose though modest style (to increase verisimilitude); and discussion of those experiments which failed.

- Each of these, however, also implied and helped to carry through a *social* technology – that is, the creation of a set of conventions for recognising and responding both to other natural philosophers and to claims about the behaviour of air. First, then, on social philosophers. These were to be modest people. If they reported experiments they were to do so in a matter-of-fact way. Whether in person or in their texts, they spoke confidently of matters of fact – matters that they had witnessed – whilst avoiding generalisations. Metaphysical speculation was out. Facts might be witnessed, but other kinds of realities were inappropriate. But, and second, they also needed to show that they were not constrained or dependent on others in any way. That they were free agents unbeholden to anyone. But what did this mean in Restoration England? The answer was very specific: only 'gentlemen' could fulfil this social requirement. Only gentlemen were not beholden to anyone else. Women, even upper-class women, were likely to be dependent on men – fathers, husbands, brothers. Their testimony was accordingly unreliable. And certainly anyone who needed to work for a living – including, not least, the technicians who built the apparatus and were physically responsible for the conduct of the experiments – was automatically rendered ineligible as a witness: in the Restoration imagination such persons were self-evidently open to being suborned.

Shapin and Schaffer talk about the way that each of these 'technologies' is embedded in and helps to carry the others.

Historically this moment in the Royal Society of London when these technologies were given shape is incredibly important because it establishes the general shape of scientific experiment, scientific witnessing, and scientific authorship with which we are still wrestling at the beginning of the twenty-first century. Leaving aside the way in which gendering and class agendas get built into the basis of scientific practice and proof right at the beginning of natural science,[110] it also establishes a very specific version of proper authorship – and a relation between authorship, authority, and accounts of reality. The scientific author is the one who witnesses, but modestly. He is the one who helps in the witnessing process by *letting the facts speak for themselves*.

But we have been there before – for this, more than three hundred years later, is what Latour and Woolgar are describing in the Salk Laboratory. Crucial to the creation of reliable statements about reality is the Othering of the personal and the subjective. If the author appears at all, it is simply as a neutral medium that passes on statements that have been produced by nature via an appropriate set of inscription devices. Nature has a specific reality. Nature speaks. The person speaks for nature, and then modestly disappears. This, as Alpers and Haraway note, is the creation of a view from nowhere.[111] And it is, perhaps historically, the first time there is the creation of a secular and naturalistic view from nowhere.

Clearly this enacts an heroic series of disentanglements, deletions and Otherings. If it is to witness reliably, then the scientific assemblage needs to detach itself from the personal and, more generally, from social interests and social context. It needs to detach itself from geographical location (scientific truths, nature as revealed, will subsequently become universal). It also needs to detach itself from specific material forms (the air-pump, properly assembled, makes possible a witness which, however, may be replicated in other forms elsewhere). All of these need to be Othered if a representational as opposed to an allegorical version of depiction is to succeed. There is, as we can still see 350 years on, nothing more damaging to a statement about reality than to say that it has been made by a specific person with specific social interests in a specific geographical location using some idiosyncratic material arrangement that cannot be reproduced elsewhere. Such a provenance is not acceptable.

So in this enactment of knowing most of the social, the geographical and the technical are rendered invisible. But this is only achieved because of a hidden set of carefully organised relations with the social, the geographical and the technical. These are the 'technologies' described by Shapin and Schaffer briefly outlined above. The method assemblage bundled together by and through Boyle – a method assemblage that is so successful – works by distinguishing between a public discursive reality and a private heterogeneity which together, and separated, secure the appearance of purity. Indeed, as we have seen, in other language this is also an argument made by philosopher and sociologist Bruno Latour (1993). Modernity, he claims, is precisely characterised by its insistence on purity – and also by its practical heterogeneity. The scientists at the Salk Institute pretend to talk – and indeed imagine that they are talking – about nature when they write their papers. But, as we have seen, in practice they are more or less precariously bundling together a heterogeneous hinterland of subsequently deleted social, material and textual resources. *Contra* appearances, nature is always entangled with culture and society. To negotiate the structure of one is to negotiate the structure of the others.

7 Imagination and narrative

The new form of critique I attempt in this book is distinguished by the framing implicit in my new story of numbers, generalizing, and certainty. This implicit set of working images/stories tells realness as emergent: what's real emerges in gradually clotting and eventually routinised collective acting, and not only human acting. I call these framing images and stories 'an imaginary,' although I hesitated in settling on this term, for it can easily be misunderstood.

(Verran 2001, 36–37)

Exploring practice

Method assemblage is the process of enacting or *crafting* bundles of ramifying relations that *condense* presence and (therefore also) generate absence by shaping, mediating and separating these. Often it is about manifesting realities out-there and depictions of those realities in-here. It is also about enacting Othernesses. If we think this way then reality, realities, take on a different significance. No longer independent, prior, definite and singular as they are usually imagined in Euro-American practice, they become, instead, interactive, remade, indefinite and multiple. But if this is right then it suggests we need ways of exploring the enactment of and the interactions between different realities. There is a need for tools that allow us to enact and depict the shape shifting implied in the interactions and interferences between different realities. There is need for assemblages that mediate and produce entities that cannot be refracted into words. There is need for procedures which re-entangle the social and the technical. There is need for the coherences (or the non-coherences) of allegory. There is a need for gathering.

The implications are profound. The cases we have looked at earlier suggest that methods in natural science and social science barely catch their own performativity and tend to disentangle themselves in theory if not in practice from multiplicity, shape shifting and the indefinite. We have seen that the predominant Euro-American mode is perspectivalist. This means that it is reductionist. It ends by authorising a single account of out-thereness. Then, in the reversal described by Latour and Woolgar (which finds its origins in the seventeenth-century circumstances portrayed by Shapin and Schaffer), and the layering described by Mol, it explains that it is the unique out-thereness that authorises the chosen narrative and necessarily disqualifies any of the possible alternatives. All these authors, but perhaps especially Mol, propose that we should undo the reductionist reversal. That nature should no longer be seen as the unique author of a single account, but something that is produced along with social and cultural arrangements. But what might such an approach look like?

Such is the topic of the present chapter, and in order to open it up I compare and contrast two very different modes of method assemblage, one drawing on and reproducing Euro-American assumptions of in-hereness and out-thereness, and the other enacting a very different version of presence and absence: that common in Australian indigenous cosmologies. I make the contrast in order to do certain kinds of work, recognising that the division flattens differences within each of the categories.

Guidebook

In the centre of Australia there is that spectacular landmark known alternatively as Uluru, and Ayers Rock. It is no coincidence that it has two names because it is at least two (and no doubt many more) realities. One of these (actually more than one) is or are Aboriginal, and the other is Euro-American. Near the beginning of the first chapter of the Australian National Park field guide to Uluru, a chapter which is called 'A Land of Extremes', we find the following:

> Why do these landscape forms exist? The geological history of this country – spanning at least 1000 million years – can help to answer this question. Indeed, thanks to the sparsity of plants and soil over much of the ground, the underlying rocks can kindle a genuine interest in geology. The rock types, colours, varying strata and the changing land forms are all very visible. Because of this the geological explanation for the landscape can be readily appreciated in the [Australian] Centre. . . . The greatest difficulty lies in comprehending how long 1000 million years really is and in visualising how the enormous changes that occurred could happen. Our lives represent such a speck in time!
>
> (Kerle 1995, 3)

Immediately after this paragraph we find the following:

> Aboriginal people have a different answer as to how these land forms came to be. For them the answers are in the Tjukurpa (djook-oor-pa) – the religious philosophy which underpins their existence. Like all religious philosophies the Tjukurpa provides explanations for the most fundamental of questions. It defines what is true, what is real and what is right. All the

land forms, all the features and all life were created during the Tjukurpa when ancestral beings travelled widely and left their marks on the surface of the earth. Nothing existed before this. This rich Aboriginal culture is evident in the Centre, both in the landscape and through the more recent celebration of Aboriginal spiritual history in stories and rock paintings.

(Kerle 1995, 3)

The second chapter of the guide, 'A Spectacular Landscape', similarly juxtaposes a Western geological account of the formation of Ayers Rock with a selection of Aboriginal stories. For instance, we learn that the monolith is composed of arkose, which is a sedimentary rock with 'small particles of pebbles of sand, quartz and feldspar with traces of iron oxides, clay, and fragments of other rocks' (Kerle 1995, 24). This rock is grey until it is oxidised by the atmosphere, when it takes on the orange-red hue so characteristic of Uluru. The ribbing which runs more or less vertically down the side of the rock and is particularly prominent on its south side is an effect of the original process of laying down the sediments. This means that the whole rock – it is over three kilometres from end to end – has been dramatically tilted by nearly 90° in the billion or so years since it was laid down. The up-ended strata from which it is composed were laid down over a long period, probably about 50 million years, and the eastern end of the rock is older. The geologists know this because they can see in these rocks what they call 'current bedding': sediments laid down in rapidly flowing currents that have a characteristic shape, because the fast-flowing water has subsequently shaved the top off the sediments previously deposited.

The guide offers a further account about how the various shallow caves round the rock at ground level were formed, and observes that:

The precise mechanism for the formation of these caves is a matter of debate between geologists. One idea is that, in places where the chemical weathering has broken through the toughened skin, the rate of weathering of the underlying arkose (which has not been toughened) is faster. Small pits become hollows and eventually caves.

(Kerle 1995, 26)

Another theory, the guide goes on to note, is that they were eroded by water held in the sand when this was at a somewhat higher level than it is at present.

This detailed account of Uluru is complemented by a geological description of a number of other topographically prominent features in the surrounding area, including the low mountains called the Olgas (which Aboriginal people call Kata Tjuta). Finally, the whole is framed by a geological-historical account of the formation of the area (which is also illustrated by a table of events), tracing more than a billion years of orogenic, erosive, tectonic, fluvial and climatic events as factors that have influenced the landscape to produce its present form.

The geological account is approximately twenty pages long. The Aboriginal account which precedes this is somewhat shorter. It starts so:

> There is no single story describing how Uluru, Kata Tjuta or any other landscape feature came into being. Anangu do not look upon Uluru as a single spiritual object. Its formation and the creation of its specific characteristics are the outcome of several stories which are not necessarily connected. Prominent features such as Uluru and Atila (Mount Corner) are regarded as an integral part of the landscape which was criss-crossed by the characters of the Tjukurpa stories.
>
> (Kerle 1995, 14)

Indeed there are many such stories. These include those of Wiyai Kutjara (the Two Boys), of Mala (the Hare Wallaby), Kuniya (the Python Woman), Mita and Lunkata (the Blue-Tongued Lizards), Tjati (the Red Lizard) and Kurpany (the Devil Dingo).[112] Here is a sample of one as it is reprinted in the guide. Tommy Manta, one of the custodians and traditional owners of the site, told the story of Wiyai Kutjara, the Two Boys, this way in 1994:

> The Two Boys came up from South Australia, and travelled towards Uluru across the south-west corner of the Northern Territory. They stopped for a while at Itarinya, a site on the Uluru side of Pirurpakalarintja, the cone-shaped peak to the west of the park. They were hunting and travelling together, and as they continued on towards Uluru, they heard the sound of the Mala at ceremonies around the rockhole that is now part of Kantju Gorge. The Mala had initially erected the Ngaltawata, their ceremonial pole at this site, but the ground was too boggy and the pole lurched sideways. They pulled it out, and replanted it in the more secure location where it still stands, turned to stone. The Two Boys travelled towards the ceremony to see what was happening. They were uninitiated boys, and had no knowledge of men's ceremonies. They were very curious.
>
> The Mala, meanwhile, were separating into their men's and women's camps to get ready for inma [rituals and song cycles] the next morning. They didn't know it, but already Kurpany was heading towards them from the west intending to destroy them. The men were resting at Mala Wati and preparing their decorations for the Inma, and the women were asleep at Tjuaktjapi.
>
> The two boys began playing in Kantju waterhole, mixing the water with the surrounding earth. They piled the mud up, getting bigger and bigger, until it was the size that Uluru is now. Then they started playing on it. They sat on the top, and slid down the south side of the mud pile on their bellies, dragging their fingers through the mud in long channels. The channels have hardened into stone, and now form the many gullies on the southern side of Uluru.
>
> (Kerle 1995, 18)

But this is (a version of) just one of the stories. For instance, the guide also tells a version of the Kuniya Tjukurpa, the Python Woman narrative. Kuniya, who travels widely across the centre of Australia, comes to Uluru from the east. After a long and exhausting journey she leaves her eggs safely at the eastern end of Uluru (the ring of eggs is visible in and makes the low rocks on the ground at that point), and she moves along the north side of Uluru, leaving serpent-like traces on the rock which are clearly visible. She hunts, but then becomes embroiled in a battle with Liru who has killed her nephew. She is furious, performs a ritual dance, drops sand in a somewhat vain attempt to control and limit the effects of her anger, and then engages in battle, killing Liru but (because she is so angry) also poisoning much of the ground round about. All of which is written into and visible on the southern slopes of Uluru: Kuniya's movements across the rock face, the sand, and the dead vegetation, all of these can be seen in the landscape. As the guide observes:

> Evidence of Kuniya's actions as she rushes towards her insulter and destroys him, is clear in the features along the Mutitjula walk. You will not just be looking at rocks and walls; you will be walking in the midst of creation and the record of events which continue today to be celebrated in story, song and ritual dance.

(Kerle 1995, 21)

Two enactments

Here, as is obvious, we have two styles of story-telling – or two very different method assemblages. The guide's sensitivity to the politics of Aboriginal–White relations reflects the special significance of Uluru both to White Australians (as a spectacular natural feature of the country's 'red centre') and to its traditional Aboriginal owners for whom, as noted above, it is a series of sites and continuing events of spiritual and spatial significance. The guide's politics of equal cosmologies also reflects the ownership status of Uluru, granted amidst much controversy back to its traditional Aboriginal owners as freehold in 1985, but (and as a part of the agreement) immediately leased back for 99 years to the Australian National Conservation Agency (together with Kata Tjuta), and open without permit to non-Aboriginal visitors under certain restrictions.[113] I will briefly consider the interaction between White and Aboriginal realities below. First, however, I want to attend to certain important differences between the two world-views – and the two sets of methods assemblages to which these correspond. And, though the guide resides primarily within a Euro-American tradition of representation, its politics of equal time juxtapose the two, and thus throw some of those differences neatly into relief.

If we start with the geological account, then we need to note that this is a popularisation. We are not here in the equivalent of the Salk Laboratory. But to call it a popularisation is to give the game away, for it draws on expert geological literatures and digests these for a wider audience. This means that its version of manifest out-thereness takes the same general form as that of the geologists. Some, indeed many, of the details that would be known to geologists are missing, but the overall narrative has the same shape. Realities are thus disentangled from and *independent* of both the Othered knower and the practices of knowing. For instance, current bedding has its own particular attributes. Further, it reveals the sequence in which originally horizontal strata were laid down. Again, over geological time it is no particular problem to summon up the orogenic forces that can tilt strata by 90°. They are known to exist in geological reality. And the as yet unresolved debates about the origins of the ground-level caves in Uluru do not erode the independence of geological out-thereness, because they are treated as a problem that will, in due course, be resolved by further investigations of reality rather than by negotiations between geologists.

Geological out-thereness also *precedes* its study – indeed in some of its features by up to a billion years. Nothing the geologists do is going to alter that reality, and the history that produced it. Whatever they learn will be a discovery. Unsurprisingly, at the same time we also discover that geological out-thereness is both quite *definite* and *singular*. A specific set of more or less complicated forces working over a billion years has produced equally specific forms of geological reality as manifested in Uluru and its surrounding landscapes. This, then, is just another instance of Latour and Woolgar's reversal, and Mol's layering. Any idea that method assemblage in geology has had anything to do with generating on the one hand a geological reality, and on the other a representation of that reality, is effaced. In the way the guide tells the geological story, it is reality which explains why one would believe this or say that about the origins and form of Uluru. A single, definite, prior and independent reality explains the statements. These have nothing to do with the social or the cultural.

Though, of course, given its implementation of a politics of equal cosmologies, this is not quite fair on the guide. (Few geological textbooks, or indeed guidebooks offer space for alternative cosmologies, except, perhaps fleetingly, as Whiggish indications of the past errors of scientists or the beliefs of natives.) It may be – but this is only a possibility – that the juxtaposition is a gathering that generates an effect of ambivalence or uncertainty. Perhaps, then, it is allegorical in effect. At any rate, before the geology in the guide there is an account of the Tjukurpa. Postponing certain difficulties (arguably it is not possible to offer an account or a narrative of the Tjukurpa within a Euro-American format such as a guidebook), what does this tell us about Aboriginal method assemblage? And in particular, what does it tell us about out-thereness in Aboriginal practice? The answer is that each of the features of out-thereness enacted in Euro-American cosmology is in greater or lesser degree undone in its Aboriginal alternative. It is also, however, that the very distinction between in-hereness and out-thereness is being undone at the same time. Let me work through the list in reverse order by starting with *singularity*.

So does Aboriginal method generate singularity? The answer is probably not: at the very best, this is uncertain. Perhaps this may be achieved in strategic, negotiated and explicit enactments of particular Tjukurpas. Perhaps it is sometimes achieved by negotiations in which the different Tjukurpas are mapped on to one another or, better metaphor, woven together to form something like a whole. Such is the basis of the semi-fictional work by Bruce Chatwin, *The Songlines*,[114] and the Uluru guide writes as follows:

> The story of Kuniya the python woman travels west to Uluru from near Erlunda. . . . If you drive to Uluru from there, the journey will take about three hours. The expectation and excitement of arriving at Uluru often means that visitors absorb very little of the country in between. This is unfortunate as that country is also part of the Tjukurpa. The Kuniya Tjukurpa – her journey, resting places and troubles – is known and sung by Anangu communities through parts of the Northern Territory, South Australia, and Western Australia.
>
> (Kerle 1995, 19)

Kuniya's journey, then, covers (and at the same time creates) much ground – hundreds if not thousands of miles. As a part of this, its narratives also belong to different tribal groups in different places, and the different stories of these groups interweave with one another to produce a kind of continuity. This, then, is a set of overlaps which one might possibly think of as a singularity.

At the same time those narratives also produce differences. The guide points to this in a phrase cited above:

> [Uluru's] formation and the creation of its specific characteristics are the outcome of several stories which are not necessarily connected.
>
> (Kerle 1995, 14)

So there are multiple narratives covering the same territories. And then there are differences, even within the Tjukurpa which attend to Kuniya, because these stories are, indeed, different in different places and are intimately and indissolubly related to those places. For instance, Kuniya and her battle with Liru belongs to, is written into, and produces parts of the south side of Uluru. The Tjukurpa – a point which we will revisit below – joins narrative and land form together in a way that cannot be dissolved. And indeed, it joins them just as seamlessly with kinship affiliation (the Tjukurpa belongs to a particular kinship group), consequent social differences, the enactment of narrative in ceremony, visual depictions, and the celebration of the sacred. All are tied up together in a more or less 'local' bundle that does not distinguish 'nature' from 'culture'. And the details of that 'local' bundle (the term 'local' does not really apply, which is why I place it in inverted commas) are likely to be known only to those with appropriate social affiliations that are determined very often not only on kinship but also gender and age-related grounds.

But there is a further consideration. Just as the Tjukurpa varies from place to place, so it is not very fixed in form at any particular location either. It is not like, say, the state opening of Parliament where the form is carefully prescribed. Instead, it varies both between different versions, and between different moments in the same place. There are several reasons for this, but one is that it is a matter for endless discussion and negotiation between those who carry it and their neighbours (Verran 1998). It simply is not fixed. More generally, postcolonial STS scholar Helen Verran describes its relative malleability in the following way (Verran is talking primarily of the Yolngu of coastal East Arnhemland who live many hundreds of miles from the Pitjantjatjara and Yakunytjatjara of Uluru, but there is little doubt that her description is also applicable to the Tjukurpa of the Western and Central Deserts):

> The knowledge of sites and their connections is contained in a large corpus of stories and the songs, dances and graphic designs which go along with the ceremonial elaboration of these stories. . . . These are performed in ceremonies where both the complex logic of gurru*t*u (the recursion of kin relations) and particular land sites are re-presented. The words of songs which celebrate this imaginary are not memorised. It is the general picture of the network of places and their interconnections that is memorised.
>
> (Verran 1998, 248)

Narratives and their enactments are not fixed in Aboriginal practice. They are negotiated and renegotiated. The fact that they are negotiable and in need of negotiation is entirely explicit. So too is the fact that those negotiations are strategic in character. The implication is that if singularity is achieved (and the extent to which this is the case is contingent and uncertain) then this is a local and momentary gathering or accomplishment, rather than something that stays in place. Aboriginal Australians, Verran suggests (2002), are theoretically multiple and practically regular – just the other way round from Euro-Americans who are practically multiple and theoretically singular.[115] So story forms recur, but there is no need for *singular* forms. Indeed, the extent to which there is anything other than a very space- and time-specific form of the *definite* is limited. Thus Verran indicates (personal communication) that if one asks a native owner about the ceremonies of a neighbouring group and place, then the inquiry is likely to be dismissed with a wave of the hand as being 'none of my business'. It is important but no comment is possible from a position of ignorance.

Singularity and definiteness, then, are uncertain. Indeed there are multiple possible realities – and indefinitenesses – but this is not experienced as a problem. At the same time, this means that the other features of out-thereness favoured in Euro-American methods assemblages are equally uncertain. So Aboriginal method, Tjukurpa, only doubtfully generates anteriority and independence. On *anteriority*, let me quote Verran again (*Wangarr* is the East Arnhemland equivalent of the term translated in English as 'dreaming'):

According to these stories, there was an eternal, simultaneous making of the people in clan groups and of meaningful foci in the land, by eternal beings as they went about their living: hunting, defecating, urinating, having coitus, menstruating, crying, and having babies. This is understood to have occurred in what is known in English as 'the dreamtime', *Wangarr* in Yolngu languages. This is often taken, incorrectly, as the far distant past, but a contrast between time as secular and eternal is probably a better way to explain it. *Wangarr* is time of a different sort (something like eternal time) to that in which we live our everyday lives (secular time); it is not time only of the far distant past. It is a time which we can find here and now, and will be able to continue to find in the future.

(Verran 1998, 247)

'An eternal, simultaneous making . . . ': this takes us into a version of time that is far from that of the geological origin story with its sedimentation, buckling and erosion. It has nothing to do with an historical-geological timeline, which operates through a linear time to produce the present. Instead, in Aboriginal cosmology the past is, as it were, continuously in the present:

Dreamings are Ancestral Beings. In that sense, they both come before, and continue to inhere in, the living generations. Their spirits are passed on to their descendants.

(Sutton 1989)

What was made is also being made now. Which adds a further point, and further helps to explain, the outrage of native owners when they were displaced as they were in the 1940s, 1950s and 1960s in a brutal policy of assimilation.[116] Removed from the sites on the land, it was no longer possible for native owners to perform the ceremonies necessary to (continuously re)create the land-and-the-person-and-the-kinship-and-the-religion-and-the-ancestral-beings. Indeed, those parts of Australia (and there are many) from which Aboriginal populations were permanently removed, whether through genocide, forced assimilation, or forced resettlement, very quickly lose their life and their form. The necessary process of eternal and simultaneous making is lost – though there are also places where that which is lost is being regenerated.

So out-thereness precedes particular enactments of that out-thereness in-here. But at the same time, the contrast doesn't really work because the world is being made and remade in each gathering, each ceremony, each re-presentation (Verran's term), each condensate. If it is definite, this is for only a moment. If it is singular, likewise. Again, if anything is anterior it is also simultaneous. And independence is similarly uncertain. In short, what is present is not strongly divided from the out-thereness it condenses. This means that Euro-American origin stories, such as that offered by the geologists, which depend on a kind of inertia in out-thereness, make little sense in Aboriginal mediations. Things are not set in place once and for all, or slowly remoulded by the operation of forces that exist out there by themselves. It is not possible to imagine that they get made, and then hold their shape as time passes. It is not possible to discover and represent them, so to speak, as some kind of operation which is separate from their existence. If they hold their shape at all it is because they are participating together in their continuing re-creation. Which means, by the same token, that out-thereness is scarcely *independent*. Land, species, naturally occurring phenomena, kinship relations, and the spiritual, all are being made together. And remade. And remade again.

So, just as there is no historical time, so there are no simple distinctions that would allow us to distinguish between an out-there that is spatially distinct from the enactments in-here. The contrast with the Euro-American accounts of historical geology are once again instructive. It seems plausible to suggest that the events which make up the latter are enacted as taking place in a four-dimensional space. Time is one of those dimensions, but the other three are Euclidean. Distances, heights and volumes, as well as the dates of various geological and topographical processes, are described. Indeed, the Uluru guide is illustrated with various isotropic and cartographic representations which show, for instance, the locations of ancient mountain ranges and alluvial fans. But, as Verran shows, Aboriginal enactment is not constituted in the same spatial idiom. It has nothing to do with area, if this is understood in a geographical manner.

This is why I placed the term 'local' in quotation marks above. The term local, it is clear, depends upon (and indeed helps to enact) a version of space as something that contains (small) localities that exist within it. This makes excellent sense in the context of the geographical and other out-therenesses of Euro-American method assemblages. What is 'local' can, then, be contrasted with phenomena that are 'global', and localities can be distinguished from one another by using Euclidean or other functionally equivalent co-ordinates. But Aboriginal methods do not work in this way. There is no global, no empty space, against which to measure and within which to locate the local. Instead Aboriginal method assemblages enact a spatiality that is indissolubly linked with the Tjukurpa, the telling, the re-enacting, and the re-crafting of the stories of the ancestral beings – events which exist, as we have seen, in an eternal simultaneous past and present. These are practices to which the notion of an empty space is foreign. To imagine an out-thereness independent of its enactments is almost literally meaningless within Aboriginal cosmology. And this is why the politics of equal time of the guide does not quite catch it: describing an Aboriginal reality out-there is already to insert it within a Euro-American metaphysical project. As, indeed, is my own account above.

Agency and dualism

Euro-American method assemblage enacts – or seeks to enact, or understands itself, as being constituted in – a reality that is independent, prior, singular and definite. Following Latour and Woolgar, but also Mol, I have argued that this is a misunderstanding. The work that makes this possible – and which also suggests that particular realities are brought into being – is systematically Othered. The uncertainty or the contingency of the realities made manifest in representations disappear. Their character, as enacted, vanishes. But in Aboriginal method assemblage those contingencies do not disappear. Here, as we have seen, everything takes effort, continuing effort. There is endless and necessary preoccupation with process. Nothing becomes autonomous. Everything has to be re-done and re-enacted. There is never closure. Aboriginal method is not, then, a process of mediation which (in its self-imagination) generates a reality that is taken to be independent, prior, and separate from the social. Unlike its Euro-American cousin it is a process of mediation that knows and recognises that this is its very nature. That knows and recognises to itself that process is inescapable. That knows that nothing is fixed. That nothing like closure is available.

There is another and complementary way of talking about this difference. This is to think of it in terms of the distributions of agency. Method assemblage in Euro-America tend to presuppose and produce a series of interrelated dualisms between the out-there and the in-here which afford the independence and anteriority – but also (to say it quickly) the *passivity* – of what is out there. These dualisms – widely discussed in the history of science – come in a number of forms. For instance, it is common to erect a division between the *human* and the *non-human*. These two classes of entities are taken to be different in kind (and there is much fuss, perhaps especially in the social sciences, if the distinction is ignored). It is similarly common to divide between knowing *subjects* on the one hand, and *objects* of knowledge on the other. Again, it is assumed that these are different in kind, and relate together in quite specific ways. In particular, it is assumed that the wise subject can 'know' the object and predict its behaviour, so long as it goes about it in the right way by disentangling itself and its methods from various illegitimate and distorting influences. This was an argument that I rehearsed in the first interlude and in Chapter 3. Again, and similarly, Euro-American method assemblage habitually distinguishes between the *social* on the one hand and the *natural* on the other. How it makes the distinction is variable, but as a rule nature is taken to be given, to be governed by general and invariant laws which determine (sometimes probabilistically) the behaviour of its components. By contrast, the social, though it may also be subject to laws of determination, in addition offers the prospect of creativity and human freedom.

These three dualisms (and others that are generally similar in form) interact and tend to reinforce one another. They do so in part because each indexes a further dichotomy. This is the divide between those classes of entities that are taken to be *active* on the one hand, and those that are known to be *passive* on the other. The human, the subject, and the social, these are or should be (mostly) active. Potentially creative, potentially discretionary, potentially autonomous – these have the capacity for action (in the standard social science sense of the term). By contrast, the non-human, the object, and nature, these are or should be (mostly) passive, acted upon, predictable. In theory how they act can be (more or less, and sometimes statistically) predicted and indeed (or so it is hoped) controlled. It is determined. Discretion and autonomy – these are not attributes that belong to the non-human. There are limits (complexity theory is about the unpredictable character of non-linear behaviour). However, in general it is taken that the natural world and its objects exhibit behaviour (in the passive, acted-upon, social science sense of the term) rather than the capacity to act. Trouble, indeed, is liable to arise when objects take off on their own and they start to show initiative. Either that, or some category error is being perpetrated: the social is not being properly distinguished from the natural. But here is the bottom line: *such patterns of dualist separation are almost entirely absent from Aboriginal method assemblages*. There is no drive to the kind of dualist division discussed by Shapin and Schaffer, and no pressure to what Latour thinks of as the purification of modernity.[117] All sorts of characters can be active, are active, are made to be active, in Aboriginal method assemblages. And this, though it is sometimes a source of trouble, is also (or so Verran suggests, and I want to follow her here) a vital source of strength.

To elaborate: Aboriginal method assemblage gathers and generates a rich plethora of actors of all kinds. Shape shifters, the Tjukurpa narratives are filled with them. The ancestral beings are part human, part animal, part natural, part social, part spiritual and part geographical. Expressed so, perhaps it is tempting to think of them as hybrids, but this isn't right either. It isn't right because these method assemblages simply don't discriminate in terms of Euro-American ontological categories in the first instance – so neither do they make hybrids between them. 'Part human, part natural', this is a description located in a Euro-American ontology with its insistence on (apparent) purity. Then again, in Aboriginal enactments, agency and intention may be (and habitually are) located in naturally occurring objects such as rocks, trees, winds, cloud formations, fire, water currents, pools, and storms. We have to go back to Shakespeare, into science fiction or, as Verran notes, into aesthetics to find analogues in Euro-American metaphysics. Such possibilities are not available in the depictions that we have of technoscience. But this is not simply the case for natural phenomena. As we have noted, animals may be agents too. Kuniya the Python Woman, Mala the Hare Wallaby, Mita the Blue-Tongued Lizard. The list is endless. But again it is necessary to go some way back – or divert into literature and perhaps especially children's literature – to find locations where agency is allocated to animals in Euro-American discourse. It is several hundred years, for instance, since animals were held legally responsible for their actions in European courts.[118]

And the point extends beyond natural phenomena and animals. In the Aboriginal world, agency is also located in ceremonies, in songs, in words, in body ornaments, and in dances. It is located in objects of technology such as motor vehicles or spears. And it is located in objects of art, such as those produced in the contemporary Western Desert tradition of painting:

> A painted design or sculpted form may . . . be considered not merely a
> human being's depiction of an ancestral Crocodile (or Kangaroo or
> Woman), but an instance of that Dreaming's manifestation in the world.
> This is why pictures and carved figures can make people sick, give them
> strength, or cause accidents to happen – or so many Aboriginal people
> believe.
>
> (Sutton 1989, 49)

Aboriginal paintings, then, or some of them, are further enactments. They are
agents. If we wanted to put this in philosophically Romantic language we
might observe that Aboriginal method assemblages generate worlds that are
enchantments: Aborigines (and their non-human allies, material and non-
material) keep up the chanting. Everywhere there is agency. Indeed, it is like
Prospero's Isle, but ten times over. And since agency is everywhere, nothing is
constructed and left to be. The universe is filled with activity. Weber's gloomy
complaints about disenchantment do not apply.

Ontological disjunction

Sometimes the worlds made in Aboriginal method assemblages detach themselves from – or are entirely apart – from those of Euro-America. Here is
Geoffrey Bardon talking of the Aboriginal artist Tim Leurah Tjapaltjarri:

> Tim often said to me that he did not really wish to know the white
> Australians, and the painting [Napperby Death Spirit Dreaming] is his
> perception of his own tribal lands and spiritual destiny in the Napperby
> cattle-station areas. He appropriates Napperby to himself as his own
> Dreaming, and by implication takes it away from its white owners.
>
> (Bardon and Tjapaltjarri n.d., 46)

'He did not really wish to know the white Australians.' This kind of separation
is visible to Euro-Americans in other places. For instance, there are four fenced-
off areas around Uluru, sites of special significance, that are prohibited to
ordinary visitors. Signs instruct those walking round the circumference of the
rock not to climb over the fences. Again, the white visitor to Kata Tjuta who
follows the elliptical curves of the road to that spectacular set of rock outcrops,
domes and valleys, discovers that he is not allowed to stop his car in most
locations along the way, and notes that he is authorised to walk on only very
restricted paths once he arrives. The latter ruling is glossed in part as a matter
of safety (the temperatures in the desert are indeed extreme in daytime except
in midwinter), but something else is going on too. This is that the visitor is
also being steered away, and avoidance is being practised:

> Kata Tjuta is a particularly important area, managed only by initiated

men. For this reason there are no Tjukurpa stories that can be told to the casual visitor.

(Kerle 1995, 16)

Space is not, as it were, isotropic: the same everywhere, essentially neutral. It is (as we have noted above) being built differently. Analogous, but less successful, is the attempt (usually more or less vain) by the traditional owners of Uluru to persuade visitors not to climb to the summit of the rock:

> Climbing Uluru . . . does provide a magnificent view and a sense of achievement, but it is against the wishes of the Aboriginal custodians because it ignores the spiritual importance of Uluru and can be dangerous.
>
> (Kerle 1995, 165)

Notwithstanding this request and the strenuous character of the climb, something like 10 per cent of those visiting Uluru indeed choose to climb the rock. How many of the remaining 90 per cent take the request of the traditional owners into account is not clear – but there is much discontent amongst tourists when, as sometimes happens after the death of a significant person, Uluru is closed for a few days. But avoidance is not simply a matter of excluding people who are white. As the citation about Kata Tjuta above and the story of the Two Boys, the Wiyai Kutjara Tjukurpa, suggest, there are restrictions on who may know about what within and between Aboriginal groups. Indeed, such is an integral part of, and an enactment of, the meshwork of Tjukurpa, the patchwork of partially connected narrative, spatial, and sacred realities that make up Aboriginal Australia. Others may know in general about the stories, and may participate in some related practices, but they will not know the full extent of the enactments and their realities.

So there are secrets, but – crucial point – these secrets and restrictions are not simply epistemological. We are not dealing here with just another, if slightly more exotic, version of the fact that (for instance) you or I don't know how to design nuclear weapons, or the size of someone's bank balance. It is not simply that some knowledge is secret or confidential. It is not that we are being refused a particular and specific perspective on certain restricted parts of a world that is common to us all. Neither is it simply that we haven't (yet?) put in the effort to master (say) the art of mass spectrometry that will (once we do so) open up parts of common scientific out-thereness that are currently closed off to us. Much more profoundly, it is that we are *not a part of these worlds at all*. Those who do not own the stories are not any part of the Tjukurpa. They do not belong to it. In a way that is very radical, and therefore somewhat difficult to appreciate from within Euro-American common sense, *we do not exist to those worlds*. Just as *they do not exist to us*.

What does this mean? One implication is that, from the point of view of the different Tjukurpas, those who are not narrated are non-people. If we exist at all, then we hardly exist. But it is important to try to get this right. It would, for instance, be wrong to imagine it as another kind of racism dressed up in some exotic, Other-centred clothes. The analogy falls because it is not a matter of reclassifying people as non-people, for instance in the same way as did the Nazis when they described the Jews as vermin (Bauman 1989). Objectionable though it may be, the Nazi method assemblage was built on and enacted its own version of Euro-American cosmology. It assumed ontological singularity or universalism – and Jews as a definite category of (non-) people existed in this cosmology. Catastrophically, they did not count as people – and as we know, the Nazis were able to enact that reality on a genocidal scale.

But in Aboriginal enactments of the world something different is going on. It is ontological universalism that is absent, rather than the denial of universal human rights. The latter, as the phrase itself reveals, depends on, and enacts ontological universalism. The problems really arise when there is interaction or interference between different particular worlds which don't have the wherewithal to recognise that they are different. For Aborigines in particular, this has happened in their disastrous encounters with Whites who, in addition to racism have also enacted Euro-American method assemblages which are committed to and presuppose universalism and singularity. As we have seen, the traditional owners indeed note with distaste that many visitors choose to ascend Uluru, while Tim Leurah Tjapaltjarri sought to avoid White Australians and reappropriate his people's land. These encounters (and worse) suggest that Euro-Americans are not entirely invisible within Aboriginal realities – perhaps seeming like ghosts or empty shells (Verran, personal communication). But from the Aboriginal world, a question which is again difficult to imagine from within Euro-American method assemblages presents itself: is communication a good? The putative answer to this question is: no, it isn't, not necessarily. Does it necessarily matter if there are enacted worlds that don't know one another? The putative answer again is no. It *may* matter if the Tjukurpa and the relevant groups overlap with one another. But if they don't, then it doesn't. *Ontological disjunction* is a possibility that might be, and indeed often is, quite appropriate.

The problem, then, is not usually within and between the enactments of different Aboriginal realities (which is not to deny that people have disputes or indeed sometimes come to blows). But this is because, to repeat the more or less inapplicable terms torn from Euro-American enactments of reality, they don't claim universalism, and whatever is enacted is specific and 'local' both to time and to place. The problem, instead, arises when Aboriginal realities overlap, as they have done for at least two hundred years, with those of Euro-America with its enactments of a passive version of spatial and temporal singularity. Leaving aside the self-evident abuses of force, the possibility of ontological disjunction is simply unavailable to the latter.

Thus as we have seen, if the traditional owners disappear and can no longer help to remake their particular worlds, then those worlds disappear. If Aboriginal children are forcibly separated from their families and their locations and removed to the Australian cities in order to enjoy the civilising benefits of white adoption (which is what happened to the 'stolen generation'), they too potentially become non-people:

> By giving each individual a personal dreaming, the community constantly recreates the ancestral world. Past re-embodiments of a single ancestor fade into the collective image of that being; it is a tenet of the religion that on death a person becomes his dreaming. To die and be buried in one's own country ensures this will occur.
>
> (Layton 1989, 15)

Genocide is irreversible, but fortunately Aboriginal practice is otherwise flexible. The problem of the disinherited child (or the visiting anthropologist who is also a non-person) can be resolved by the simple process of adoption – at which point the adoptee becomes real, is practised as real, and is able to participate in and carry the narratives and the realities of the relevant Tjukurpa.

Recognising enactment

Here is the contrast.

Euro-American method assemblage manifests a world in its depictions that is ontologically single, and therefore inhabited by a finally limited number of objects, forces and processes that may be more or less well known. Like the space–time boxes of the geologists. That which is not clear is at least in principle susceptible to clarification. Inquiry thus involves delineating those entities, or correcting misapprehensions about them. The assumption is that final agreement can and should be reached at least in principle (though subsequent corrections may become necessary if error is discovered). As a part of this, the possibility of a practice for knowing which recognises that entities are being endlessly enacted and (as a part of this) are being differently enacted in different locations and in different contexts, is repressed. So this is the tension. *In the midst of representational singularity there is multiplicity.* But this is not seen. The multiple or the fractional, the elusive, the vague, the partial and the fluid are being displaced into Otherness. Necessary, indeed enacted, but Other. Instead what comes into view is a reality out-there that is independent, prior, single and determinate. To the extent that it is important, it is the job of the investigator to try to determine the character of that reality. Once this happens arbitration is possible: different perspectives can be compared, and the correct solution determined. As we have seen, this is *a solution that denies the possibility of an explicit ontological politics.* It enacts such a politics, yes, but it does not see that it is doing so – a benefit, if that is what it is, only possible, as Verran observes, from a location of privilege.

All of this is in contrast with Aboriginal method assemblage. As we have seen, this is capable of enacting an ontological multiplicity that comes close to ontological disjunction. It achieves this because there is no universal or general, and instead everything is relatively specific, relatively 'local', enacted at particular places on particular occasions. Because there is no overall privilege. This means that that which is not clear is not necessarily waiting to be made clear. Perhaps it is diffuse, of marginal concern, and therefore hardly exists and can be left indefinite. Perhaps, however, it is important, in which case it becomes a topic for discussion and negotiation. Verran:

> Aboriginal Australian peoples generally understand themselves as having a vast repertoire by which the world can be re-imagined, and in being re-imagined be re-made. In English this usually goes under the title of 'the dreaming'. I think a more helpful name for this conceptual resource is 'the ontic/epistemic imaginary' of Aboriginal knowledge systems. It is this imaginary, celebrated, venerated and providing possibilities for rich intellectual exchange amongst all participants in Aboriginal community life, which in part enables the eternal struggle to reconcile the many local knowledges which constitute Aboriginal knowledge systems. Many Aboriginal communities know how to negotiate over ontic categories.
>
> (Verran 1998, 242)

Verran thinks of 'the dreaming' (Tjukurpa in the Western Desert and the *Wangarr* for the Yolngu) as an 'ontic/epistemic imaginary' because it is a rich cultural resource for, and an outcome of, (re)telling and (re)making realities.[119] Indeed, it is just the kind of resource a group would need if it were serious about its ontological politics (Verran prefers to talk of 'ontic' and 'epistemic' politics rather than of ontology and epistemology because the latter terms tend to imply stable, fixed, and 'philosophical' systems). But (though using my terminology), she is also saying that a group will be serious about its ontological politics and serious about its imaginaries if it is also serious about *negotiating* with other realities, cultures and groups about what there is. If it wants to remake its imaginaries. If it is willing to remake its imaginaries and let them settle into novel forms. Let them, as she puts it, clot in new ways. The implication is that politics gets mixed up visibly with reality-making, and also with the use of metaphors from the 'imaginary'. Verran again:

> discussions are likely to be tied up with the ongoing struggle for cognitive authority, waged through pitting metaphor against metaphor. There is often heated, and overt struggle, over whose metaphor is going to prevail. Given time one metaphor will carry the day, and it will have been greatly enriched by the controversy surrounding its being settled upon.
>
> (Verran 1998, 242)

Realities, then, get settled through an explicit negotiation about metaphors for telling and metaphors for being – though they are only settled for the time being. Other metaphors – and so other partially related putative realities – are waiting in the wings, and next time they will appear again in the process of negotiation.

All of which suggests another resource for starting to undo the refusal to think about practical enactment in Euro-American method. This is that *we keep the metaphors of reality-making open*, rather than allowing a small subset of them to naturalise themselves and die in a closed, singular, and passive version of out-thereness. That we refuse the distinction between the literal and the metaphorical (as various philosophers of science have noted, the literal is always 'dead' metaphor, a metaphor that is no longer seen as such). That we refuse the dualism between the real and the unreal, between realities and fictions, thinking, instead, in terms of *degrees* of enacted reality, or more reals and less reals. That we seek practices which might re-work imaginaries. That we work allegorically. That we imagine coherence without consistency.

And this is not just a matter of theory, but applies in Australia to the politics of land ownership. The story is a large one,[120] but in outline, a long-standing Australian law to the effect that Australia was uninhabited when the English arrived was overturned in the 1990s. Under certain circumstances, 'traditional ownership' was recognised – and with this the need for White pastoralists, conservationists and mineral companies, and Aboriginal owners to negotiate about land title. But how to reconcile the two approaches and their different notions of land?

For pastoralists and other Euro-Americans, land ownership rests on and is performed by legal documents which rest on the enactments of survey methods and cartographic method assemblages which in turn condense and enact a spatial reality consistent with Euro-American expectations. We know, by now, what this looks like: space exists, it is out there, intrinsically empty and isotropic, definite, prior and independent. But Aboriginal methods for knowing and making the land are quite different. However, with the Native Title Act, pastoralists and other white people *have* to sit down with the Aborigines and find some way of relating these two traditions and two realities together. And this has been – and is being – worked out in a variety of circumstances, for instance in conversation.[121] Verran's overview of this process ends by suggesting that in these new circumstances pastoralists, and by implication other Euro-Americans, can no longer hold on to the limited reality proposed by the closures of cartography. They will need instead, she suggests, to embrace some, at least, of the skills in epistemic and ontic negotiation of their Aboriginal interlocutors. Big and painful changes, for sure, which will lead to a world of less certainty. But a world in which the politics of ontology is no longer practised by stealth.

INTERLUDE:
Hinterland and reality

In this book I have used a range of metaphors for talking about the 'out-there'. These have included: hinterland; manifest absence; absence as Othered, fluxes, relations, and resonances. I have avoided using one of the most common terms in the social science literature: that of structure. I hope the reason for this is clear. The idea of 'structure' usually implies not simply a generic or primitive version of out-thereness, but additional commitments to independence, anteriority, singularity and definiteness. To talk of 'structure', then, is probably to imply that the real is out-there, in definite form, waiting to be discovered – even if there are major technical difficulties standing in the way of its discovery in practice.

Assumptions of this kind underpin contemporary versions of realism.[122] The latter argue that scientific experiments make no sense if there is no reality independent of the actions of scientists: an independent reality is one of the conditions of possibility for experimentation. The job of the investigator is to experiment in order to make and test hypotheses about the mechanisms that underlie or make up reality. Since science is conducted within specific social and cultural circumstances, the models and metaphors used to generate fallible claims are, of course, socially contexted, and always revisable.[123] Nevertheless the assumption is that out-thereness is independent and definite. Different 'paradigms' relate to (possibly different parts of) the same world.

The metaphysics that I have been exploring are also realist, but only in the primitive or originary sense. They assume general flux of out-thereness but nothing more. The position is close to that of Ian Hacking:

> There is only one way in which my thesis is contrary to a bundle of metaphysical doctrines loosely labelled 'realist'. Realists commonly suppose that the ultimate aim or ideal of science is 'the one true theory about the universe.' I have never believed that even makes sense.
>
> (Hacking 1992, 31)

For Hacking:

> Our preserved theories and the world fit together so snugly less because we have found out how the world is than because we have tailored each to the other.
>
> (Hacking 1992, 3)

Hacking has a constructivist view of scientific experiment. In the particular world in which we happen to live, scientific inquiry has, as a matter of fact, arrived at a set of particular conclusions, and created an empirical reality to match. To arrive at the version of method assemblage argued in this book we need to move from a focus on construction to attend to enactment. This, as we have seen, allows or requires us to add in multiplicity. The possibility – indeed the likelihood

– that tensions appear between different enactments (and knowledges) of reality is made manifest. And with this possibility that realities may be crafted and acknowledged as indefinite gatherings, coherent or non-coherent, ambivalent, allegorical, and within or beyond language. If we wanted to play games with words we might think of calling this position *enactment realism*.

Of course if all the social and sociological method assemblages were co-ordinated in a single place, if they were all brought together, then such an enactment might craft a more or less singular, and who knows, definite hinterland or 'structure' across a range of different locations. It would seem coherent. But this ignores everything that we have learned about multiplicity and multiple enactments. Even more importantly, it also ignores the distinction we have made between manifest absences on the one hand, and Othering on the other. For if realities are multiple or fractional, then the resonances and the patterns of the different absences made manifest overlap – but only fractionally. And if there is always absence as Othering, then we can say nothing in general about that Othering. Only in particular, and at particular moments.

In the end this is a matter of metaphysics. If we feel uncomfortable without clear, definite and singular accounts of clear, definite and singular structures, then that is how it is. However, if we are able and willing to tolerate the uncertainties and the specificities of enactment, flux and resonance, then we find that we are confronted with a quite different set of important puzzles about the nature of the real and how to intervene in it. Perhaps, for instance, the 'great structures' of inequality are to be understood not as great structures but as relatively non-coherent enactments which nevertheless resonate or interfere with one another to keep each other in place. Latour offers a parable about colonialism written in such terms. The whites who arrived in the colonies, he says, were a rabble. But:

> They were stronger than the strongest because they arrived *together*. No, better than that. They arrived *separately, each in his place and each with his purity*, like another plague on Egypt.
>
> The priests spoke *only* of the bible, and to this and this alone they attributed the success of their mission. The administrators, with their rules and regulations, attributed their success to their country's civilizing mission. The geographers spoke *only* of science and its advance, The merchants attributed all the virtues of their art to gold, to trade, and to the London Stock Exchange. The soldiers simply obeyed orders and interpreted everything they did in terms of the fatherland. The engineers attributed the efficacy of their machines to progress.
>
> (Latour 1988, 202)

Perhaps this particular gathering is helpful. Perhaps it is not. But as a style it deserves some thought. And, to be sure, we have visited the possibility already in Chapter 6 in our discussion of ambivalence and allegory. To put it differently, this is the idea that what engineers call 'loosely coupled systems' are more robust than those that display a single and definite logic. Perhaps out-thereness resonates to produce dramatic patterns without single and definite structures at all. Perhaps things hold together precisely because they don't. Citations: the bush-pump; the cervical screening programme; the Quaker meeting for worship. [124] In which case the great inequalities and distributions might be better understood not as structures but as non-structures. And we will need, for better or for worse, to find ways of exploring the partial overlaps of hinterlands, the manifest absences, if we want to get to grips with those inequalities.

8 Conclusion: ontological politics and after

Introduction

This book is an account of the state of method. The argument has been that method in social science (and natural science too) is enacted in a set of nineteenth- or even seventeenth-century Euro-American blinkers. This means that it misunderstands and misrepresents itself. Method is not, I have argued, a more or less successful set of procedures for reporting on a given reality. Rather it is performative. It helps to produce realities. It does not do so freely and at whim. There is a hinterland of realities, of manifest absences and Othernesses, resonances and patterns of one kind or another, already being enacted, and it cannot ignore these. At the same time, however, it is also creative. It re-works and re-bundles these and as it does so re-crafts realities and creates new versions of the world. It makes new signals and new resonances, new manifestations and new concealments, and it does so continuously. Enactments and the realities that they produce do not automatically stay in place. Instead they are made, and remade. This means that they can, at least in principle, be remade in other ways.

The consequence is that method is not, and could never be, innocent or purely technical. If it is a set of moralisms, then these are not warranted by a reality that is fixed and given, for method does not 'report' on something that is already there. Instead, in one way or another, it makes things more or less different. The issue becomes how to make things different, and what to make. Within the (always to be tested) limits of the resonating hinterlands of the currently performed patterns of realities there are different possibilities. Method, then, unavoidably produces not only truths and non-truths, realities and non-realities, presences and absences, but also arrangements with political implications. It crafts arrangements and gatherings of things – and accounts of the arrangements of those things – that could have been otherwise. But how to think this? How to move away from the idea that method is a technical (or moralising) set of procedures that need to be got right in a particular way? How to move from the legislations that we usually find in the textbooks on method? Away from the completed and closed accounts of method? Away from smooth Euro-American metaphysical certainties?

In this book I have tried to develop a set of vocabularies for thinking about method, its operations, and its performativity. Following authors in the history, philosophy and sociology of science, I have widened the notion of 'method' to include not only what is present in the form of texts and their production, but also their hinterlands and hidden supports. To catch this process of crafting and bundling I have proposed the notion of method assemblage. The argument is that method is not just what is learned in textbooks and the lecture hall, or practised in ethnography, survey research, geological field trips, or at laboratory benches. Even in these formal settings it also ramifies out into and resonates with materially and discursively heterogeneous relations which are, for the most part, invisible to the methodologist. And method, in any case, is also found outside such settings. So method is always much more than its formal accounts suggest.

There is a more formal way of putting this which is to say that method assemblage is a continuing process of *crafting and enacting necessary boundaries between presence, manifest absence and Otherness*. This form of words borrows from the post-structuralist insight that making anything present implies that other but related things are simultaneously being made absent, pushed from view, that *presence* is impossible without *absence*. Thus representations go along with something out-there to represent – and a lot more besides. The same is also the case for objects, which are crafted with a context out-there with which they interact more or less indirectly. This, then, means that method assemblage makes something present by making absence. Formally I treat it as the enactment of *presence, manifest absence*, and *absence as Otherness*.[125] More specifically, it is the crafting, bundling, or gathering of relations in three parts: (a) whatever is in-here or *present* (for instance a representation or an object); (b) whatever is absent but also *manifest* (it can be seen, is described, is manifestly relevant to presence); and (c) whatever is absent but is *Other* because, while necessary to presence, it is also hidden, repressed or uninteresting. The issue, then, becomes one of imagining – or describing – possible ways of crafting method, obvious and otherwise.

I have also argued that method assemblage can be understood as *resonance*. This is because it works by *detecting and creating periodicities in the world*. The picture of reality that lies behind this removes us from the most common version of Euro-American metaphysics – the sense that the real is relatively stable, determinate, and therefore knowable and predictable. The alternative metaphysics assumes out-thereness to be overwhelming, excessive, energetic, a set of undecided potentialities, and an ultimately undecidable flux. Sometimes, however, and in method assemblage, out-thereness crystallises into particular forms or (a different metaphor) collapses for a moment into decidability. If method assemblage can be seen as resonance then this is because it detects all the periodicities, patterns or waveforms in the flux, but attends to, amplifies, and retransmits only a few whilst silencing the others. The question is: what does standard method assemblage silence? Which possible realities does it refuse to enact in its dominant insistence on that which is smooth? And how might it be crafted differently?

Realities

The largest part of the book is a survey of the character of those possible realities. I have suggested that dominant Euro-American enactments produce and presuppose forms of manifest absence that are *independent* and *prior* to an observer; *definite* in shape and form; and also *singular* (there is only one reality). Along the way I have also noted that Euro-American method assemblage usually assumes *constancy* (there are general and invariant laws and processes, and nothing changes unless it is caused to change), *passivity* in the objects that it discovers (they stay the same until they are caused to change) and *universality* (what is absent is generally the same in all possible locations).

All this is self-evident in Euro-American metaphysics, but attending to the *practice* of its methods reveals, first, that these assumptions are systematically breached, and, second, that the fact that this is happening is repressed or displaced into Otherness. Dependence and simultaneity exist instead of (or alongside) independence and anteriority. Mol's studies of hospital realities suggest that objects that are singular in theory are multiple or fractional in practice.[126] Object constancy is similarly enacted – and breached. As, too, is universalism. (If there are multiple realities then there are no universals, only the appearance of universals.) In addition, the assumption of definiteness is also violated. Methods, construed in the standard way, are usually committed to clarity and often to precision. But since method assemblage ramifies out into the patterning resonances of a wide hinterland, this includes gatherings that are manifestly allegorical, ambiguous, indefinite, unclear or tacit. And finally, it appears that passivity is only achieved because the active process of producing realities is pressed into Othered absence and the dualist reversals discussed in the last chapter are enacted. That out-there is made into a domain that seems quite removed from what is in-here.

The suggestion is that the realities enacted in Euro-American method assemblages are complex, but also that most aspects of that complexity are denied. It may be that this Othering has its merits. I have noted, for instance, that Latour (1993) insists that (non-)modernity flourishes because it makes complex hybrids, and that this is easy to do, precisely because it *is* also denied. For Latour, then, though the smoothnesses of purity reveal self-misunderstanding, they are also a good. So perhaps there are advantages to what he calls the 'non-modern' constitution, but there are also difficulties that follow from this and related denials. To talk about these it is convenient to consider versions of representational presence, forms of depiction.

Gatherings

Some modes of method assemblage produce conventionally acceptable statements, representations, or depictions of the realities for which they stand. But terms such as 'statement' or 'representation' are specific. This is why I have talked, instead, of presences and gatherings. My aim is to be permissive, and to say nothing either about the appropriate shape, or the materiality, of whatever is crafted into presence. All that is being said is that matters are relational: what is being made and gathered is in a mediated relation with whatever is absent, manifesting a part while Othering most of it. Much of the book has been a survey of the materialities and shapes of possible presences. My interest has been to extend the list beyond those that are normally taken to be appropriate in common understandings of method. Thus the list of depictions has included the following: *texts*, for instance medical textbooks, ethnographies, scientific papers, spreadsheets, and the traces generated by inscription devices;[127] *visual depictions*, for instance, photographs of angiographic X-rays, cross-sections of blood vessels, chromatographic separations, or Aboriginal artwork; *maps* of various kinds, including but not limited to those generated by Euro-American cartographic and survey methods;[128] *human apprehensions*, some of which are conventionally understood to be relevant to method, as with the visual skills of scientists, and some of which are less conventional. Examples have included the sense of disorder experienced by researchers on a visit to an alcohol treatment centre, the sense of horror of those who witnessed the scene of the Ladbroke Grove collision, or the apprehension of spiritual realities in Quaker worship; *bodies*, as in the Ladbroke Grove crash or, one might add, in the physical condition of those suffering from alcohol poisoning or the poor skin condition of those with severe lower-limb atherosclerosis; *machines*, for instance, in the form of inscription devices (which can, as we have seen, be understood as routinised statements), but also in the form of devices that do not (primarily) work to produce traces. Examples have included the bush-pump, and, very differently, the wreckage of the Ladbroke Grove railway accident; *ceremonies*, for instance those of the Kata Tjuta Aborigines (that have not been described here because we do not belong to them and do not know about them), or the Quaker meeting for worship; *demonstrations*, as in the theatre of proof mounted by Robert Boyle, and described by Shapin and Schaffer; *conversations*, like those described by Mol in the consulting room or in the operating theatre; and *allegories*, which I have argued are ubiquitous, but sometimes, on the other hand, also recognise their character as allegorical.

The list is not exhaustive. It is very important that it *not* be seen as exhaustive. Other possibilities, more or less conventional, that come readily to mind include: musical performances; surgery; sport; physical lovemaking; games; model-making; architectures; cities; films and documentaries; prayer; physical exercise; collages and pin-boards; dance; masque; driving; cooking; flânerie; sculpture; natural phenomena of all kinds; gardens; and landscapes. And no doubt a lot more besides. These, then, are all crafted forms of presence. They do not *have* to be understood as allegorical methods of depiction for they also work in other ways, or have other roles. But my point is that it is *possible* to treat them that way. And the character of the list is revealing. It shows us that research methods as conventionally construed in natural and social science are limited in two important respects. First, they are materially restricted. The idea, for instance, that a garden or a religious ceremony or a game or a meal might be an allegory for, resonate with, and help to craft a particular reality, though just about recognisable from common sense (and a commonplace in an anthropology of symbolism), lies far beyond the limits proposed in standard method. Second, they are also limited because they tend to create and make manifest absences that are taken to be independent, prior, singular, definite and passive and all the rest.

We need to be cautious. There is no particular correlation between material forms of presence and the absences to which these relate. Both are made in mediation, and the argument is not reductionist. In any case, as we have seen, Euro-American method depends like any other on Othered entities and relations that it cannot make manifest. Othering is inescapable. Even so, the limited materialities of standard methods restrict the extent to which other realities can be enacted in at least two ways.

First, certain kinds of realities are condensed at best with difficulty into textual or pictorial forms. For instance, mystical spiritual experience cannot be captured in words. It is, precisely, excessive to the word and can only be gestured at textually. Quaker and Aboriginal lives suggest that spiritual experience also needs to be caught in bodily experiences, or apprehensions, or dance, or in art. Narrative that represents a reality goes only so far. But the argument is not simply important in the context of the spiritual. Many other realities are like this too. Is it possible to *describe* emotional ecstasy, or love, or pain, or grief, or fear? Scarry argues that language is other to pain (Scarry 1985). At best words may point to it ('a stabbing pain'). So here the condensate comes primarily in other forms. The body in pain. Or a piece of music ('our tune'), or a landscape, or bodily actions, or the sight of a loved child. Many realities craft themselves into materials other than – or as well as – the linguistic.

It may, of course, be argued that while love or pain or religious experience are realities, they are not the kinds of realities relevant to social science. The argument deserves attention. There are good reasons for holding religious experience or love separate from academic or policy inquiry. I'll return to this briefly below, though I take it that the argument is contestable. But, in any case, even realities more conventionally relevant to natural and social science are excluded by their dominant methodological practices. We have encountered a number of examples above: the organisation of health care for alcoholic liver disease, the character of lower-limb atherosclerosis. We can catch the argument so: if matters are non-coherent, then to try to *describe* them as non-coherent may miss the point since it insists on generating a form of coherence. Some other allegorical mode might be better. Some other kind of gathering. One that stutters and stops, that is more generous, that is quieter and less verbal.[129]

Second, even within the domain of texts and other inscriptions, academic method assemblage also sets limits to proper form. Some (the article, the research report, the grant application, the review, the book, the seminar) are permissible. So, too, are certain kinds of maps, diagrams, graphs, and photographs. But many forms of text and visualisation are not. On the whole, for instance, academic method assemblage does not condense in the form of poetry, fiction or theatre (there are, of course, exceptions). Few visual depictions in research follow the conventions of fine art or comic strips or film or advertising. Many textually or otherwise inscribed realities, then, are being ruled out. Academic texts are usually read as more or less technically adequate descriptions of external realities. Unlike novels they are rarely read for themselves. And, though there are exceptions, neither are they commonly read as resonating participants in the enactment of the realities that they also describe.

These restrictions have their place. They make it possible to produce particular realities: presences that (are taken to) describe, mirror, correspond or work in relation to specific and singular realities. Shapin and Shaffer describe the origins of the Euro-American attempt to make and tell the world this way. But the result has been to displace or to repress methodologies and realities that make and depict the world differently. In Euro-America the inscriptions that condense *ontic/epistemic imaginaries* belong to the novel or to poetry or to art and not to serious research method. As do those that condense *non-coherences* (James Joyce?), *overpowering fluxes* (Edvard Munch?), *indefinitenesses* (Mark Rothko? Franz Schubert?), *multiplicities* (Georges Braque?) or *fractionalities* (Steve Reich?). Perhaps all this is fine, representing *inter alia* (as Helen Verran (1998) observes) a modernist division of labour between truth and the aesthetic. On the other hand, it is also costly. It is costly since it Others imaginaries, fluxes, indefinitenesses and multiplicities – even as it draws on them. And, at the same time, it denies the various desirable effects – the various goods – that these might carry and enact.

Goods

So what are the 'goods' that method assemblages might generate? In which they might participate?

I have discussed two goods at length above: *truth* and *politics*. If methods are performative they discriminate by trying to enact realities into and out of being. But as we have seen, though this is usually displaced into Otherness, they also enact different realities in different places and on different occasions. This means, as again we have seen, that truth is no longer the only arbiter. No longer, let me stress this, the *only* arbiter. For it is still very important, crucially important, in many crafts. 'Is this true?' Yes, this remains a critical question, not one that will go away. It has been a continuing theme throughout this book that method assemblage does not work on the basis of whim or volition. It needs to resonate in and through an extended and materially heterogeneous set of patterned relations if it is to manifest a reality and a presence that relates to that reality. So *truth is a good*. It remains a good. Method assemblages that do not produce presences that have to do with truths may be attractive, there may be other reasons for generating them, but whatever they are, they are not about the out-therenesses of possible realities.

But truth is not the only good. Enter, then, *politics* which is a second good in this mode of listing. If politics is about better social (and now, we learn) non-social arrangements, and about the struggles to achieve these, then method assemblage and its products can also be judged politically. It does politics, and it is not innocent. In its different versions it operates to make certain (political) arrangements more probable, stronger, more real, whilst eroding others and making them less real. This, indeed, is one of the reasons why I, in common with many scholars in STS, feminism, and cultural studies, would like to open up and broaden the standard reality-setting agendas of Euro-American technoscience. It may or may not be a political good to create (for instance) multiple ontic/epistemic imaginaries. Whether this is so will depend on the circumstances, on the content of those imaginaries, and where one is oneself located. However to propose a blanket prohibition of imaginaries in the method assemblages of truth-making (for instance by exiling such imaginaries to the peripheral realm of aesthetics) is not a good. It is a politics of Othering which presupposes and enforces the dictum that singularity is destiny, that disenchantment is in the nature of things, and that multiplicity is a mistake.

Similar arguments apply to definiteness. Whether realities that are fluid, fractional, multiple, indefinite and active are good or not has to be judged circumstance by circumstance. There is no general rule. These are not political goods in and of themselves. (Compare a train crash with a bush-pump.) But to enact general prohibitions on (the recognition of) realities that display these attributes is to enact a class-politics of ontology that *is* a bad. Greater permeability and recognition of fluidity and all the rest, overall this cannot be a bad. This, then, is the end of political innocence. Truths are not, as the theory of ideology tended to suggest, necessarily in conflict with politics. Truths and politics go together one way or another. Or at least they may go together. And once the performativity of method is recognised this implies responsibilities to both of these goods.

This is an argument that is recognised, albeit in a somewhat different idiom, in a variety of politically radical interventions in contemporary technoscience: for instance in some versions of feminist writing.[130] But there are other goods too, and sometimes these get lost in the preoccupation with truths and politics. Indeed, we have tripped over one above: that of *aesthetics*. Thus talk of 'beauty', then, or 'elegance', or 'fit', or 'economy' indexes a further set of goods. Again some care is needed. What counts as beauty can neither be determined in general, nor out of context. (Absolutist theories of aesthetics are no better guide to tastes-in-practice than are those of epistemology to truths-in-practice.) But where do aesthetics turn up in an explicit manner in practice? The answer is, not very often in those forms of method assemblage that have to do primarily with the enactment of truths. And, interestingly, to the extent that they do, they turn up more in the exact sciences than in social science. Mathematicians often talk of elegance (though mathematics, of all the science-related disciplines, is the one that most celebrates imaginaries), and similar concerns can be found, for instance, in physics.[131] But, at least at first sight, the idea that social science truth might somehow be related to beauty seems improbable: the proliferation of more or less ugly jargon seems to be more common. However, overall, in the current arrangement of goods, aesthetics have relatively little to do with truths, social scientific or scientific. Mostly they are delegated to the arts or to consumption. We witness, then, a further refraction of the modern division of labour that separates out the different domains and acts to protect truth from other goods.

But if the argument developed in this book is sustainable, then it is not obvious that this division of labour is a good. It is not that, as if in some contemporary version of fascism, it becomes a self-evident good to celebrate the aesthetic before all else. Beauties will need to live alongside truths, and alongside politics too. As I have noted above, they are, in any case, multiple in their enactments and forms. But their blanket absence from the processes of crafting realities is not a good. It works to exclude ontic/epistemic aesthetic imaginaries. It represses their fluidities, fractionalities and indefinitenesses. And it denies us any grounds for negotiating to enact realities that are true and politically desirable but are *also* beautiful. In short, it denies to reality-making any responsibility for beauty, treating this instead as a category error. Implicitly, then, ugliness is okay so long as whatever is enacted is true. Again there is no general rule. This may sometimes be okay. But it is not necessary to insist that the aesthetic should always be collapsed into the epistemological to argue that the extent of their current separation is a bad.

The divisions of labour and the prohibitions and separations that accompany it reach further. Perhaps *justice* can be elided with politics – or perhaps not. But if not, then this is another good that is rigorously excluded from the enactment of truths. However, if this is the case, then questions similar to those rehearsed above crowd in. Perhaps, for instance, it is worth considering whether some realities are more just than others? Or whether partial-realities that are more just could be rendered more real than they actually are? And similar arguments apply to further versions of the good. For instance, is the *spiritual* a good? Quakers and Australian Aborigines are not the only people in the world who live in a world permeated by the spiritual, and who participate in assemblages that enact its realities. Who know (let us broaden the category) that the world is charged and run through with moments of inspiration. The spiritual and the material – here these cannot be distinguished. But if we note this, and note that inspirations or spiritualities are, or may be, enacted in some worlds, then arguments like those above need to be worked through here too.

First, to imagine that inspirations are real, that they are a good, and that they are relevant goods to living in the world, is not to insist that they are the only goods. (Spiritual reductionism leads to religious fundamentalism which does not seem to be much of a good except, perhaps, to those who try to enact it.) Second, to imagine that inspirations are enacted realities is not to say that specific versions of the spiritual or of inspiration amount to a good in any particular context or practice. There is no general rule. (Fascism works in part through inspiration and charisma.) Third, to suggest that they are goods that may be relevant to the enactment of truths is not to say that they would, or should, always be so enacted. Such are the kinds of cautionary notes that I need to sound. But their blanket displacement also incurs costs. For instance, spiritualities or inspirations can be understood as manifestations of ontic/epistemic imaginaries. The implication is that at least in their less codified and power-saturated forms they may be capable of making versions of the real that distribute agency more generously and less parsimoniously, allocating it in a manner that is less dualist, less prone to treating the natural as passive, reacted upon, brute.[132]

And if we escape the brute singularity of the world, the sense that reality is destiny? Then there will be a need to weave together different goods. Perhaps there will be the need to imagine and practise world-making as flows, vortices, or spirals in which links between different partially connected goods are made and remade. In which truths and spiritualities and inspirations and politics and justices and aesthetics are variously woven together and condensed at particular moments, and partially separated at others. A choreography, a dance, a process of weaving, of partial connection and partial separation, which might then spill over too into the last great category excluded by the divisions of labour of modernism, that of the personal, the emotional, the realm of fears and loves and passions.

Haraway notoriously observes that she 'would rather be a cyborg than a goddess'. That is fine, but if we think of method assemblage and goods then this suggests that it is not always necessary to make a choice.[133] For there are different goods. But none is entirely separated from the truths of reality, except in convention, in the modernist settlement, in the forbidding conventions of method.

Re-ordering

I started with the desire to subvert – or at least to raise questions about – current social science methods. Current methods, I argued, have many strengths, but they are also blinkered. Along the way I have tried to show that they both presuppose and enact a specific set of metaphysical assumptions assumptions that can and (or so I suggest) should be eroded. But what of practice? What might alternative methods look like in practice? What would it be to practise methods that were slow, uncertain, that stuttered to the stop, the attention to process, proposed by Appelbaum? What would it be to practise quiet method? Method with fewer guarantees? Method less caught up in a logic of means and ends? Method that was more generous?

The answer, of course, is that there *is* no single answer. There *could* be no single answer. And, indeed, it is also that the ability to pose the questions is at least as important as any particular answers we might come up with. So if the arguments developed in this book make it possible to debate a wider range of methodologically relevant questions, then I will be happy. So what are the kinds of issues we might debate? Here are some of the more obvious possibilities:

1   *Process*. In Euro-American method the bias is against process and in favour of product. Look at any grant application form and you will see that the rules of method are imagined as a means to an end for knowing better or intervening. The practicalities of knowing are bracketed and treated as technique. So the first set of methodological questions has to do with the analysis of practice. Can we, and should we, be looking for ways of attending to the practicalities of making realities? Of attending to the mediations of method assemblage? Of exploring the various ways in which they generate realities on the one hand, and condensates on the other? And if the answer to these questions is yes, then how might we do this? What are the modalities for praxiography?

   This, then, is the first root question: should we have a *concern with ontological process*? The answer I have offered in this book is that this is important, indeed vital. Means/ends divisions cut the cake in a particular way. Parts of process, enactment, can be pushed into a means/ends scheme, but other parts cannot. To understand the continuing and uncertain enactment of ontology and to craft it well, we will need to treat with the uncertainties and undecidabilities of process as well as with means and ends.

2   *Symmetry*. Euro-American approaches to method tend to set up rules for discovering realities. These rules distinguish between good and bad method. They tell, for instance, how results should be acquired, and the proper ways in which they should be reported. This is a kind of asymmetry. Bad results are derived from bad methods, and good from different and acceptable forms of method. So a second set of methodological questions has to do with symmetry. Can we, and should we, consider *all* practices for producing realities and condensates as possibly appropriate methods? Can we and should we be more generous in our definitions of method? Should we stop ruling whole classes of practices out of court? I have argued that we should: that there are many possibly appropriate methods. I have also argued that if we want to understand our methods then we need to treat them symmetrically, to explore them without, in the first instance, judging their adequacy in terms of our prior assumptions about what is methodologically right and what does not pass muster.[134]

3   *Multiplicity*. If we focus on practice then we are led to multiplicity since there are many practices crafting many realities. Truth is no longer the only arbiter and reality is no longer destiny. There are (to put it too simply) choices to be made between the desirability of different realities. The world could always be otherwise. Can we cope with this? My answer has been yes. If realities are being enacted multiply, then I have argued that it becomes important to think through modes of crafting that let us apprehend that multiplicity. We need ways of knowing about and enacting fractionality or partial connection.

4   *Reflexivity*. If we attend to practice we are also led to issues of reflexivity.
    In particular, we need to ask whether we are able and willing to recognise
    that our methods also *craft* realities. I have argued that it is both possible
    and important to do this, and that this it not self-indulgent but necessary
    in a world of multiple 'goods'. But how to do this? One answer is that we
    need, as I have noted above, to attend to process. In particular, however,
    I have suggested that we might attend to the way in which method enacts
    divisions between different forms of absence: absence made manifest, and
    absence as Othering. How that boundary is made and remade, this
    becomes a central concern. As does the related issue of our own unavoidable
    complicity in reality-making.

5   *Goods*. The focus on practice and the commitments to symmetry, multi-
    plicity and reflexivity together suggest that truth is no longer the final
    arbiter. But if this is right, then there are other goods to be taken into
    account. The question, then, is are we able and willing to recognise the
    multiplicity of goods enacted in method assemblage? I have argued that
    this is vital. But how should we think about different goods? How might
    they be enacted and related, and where? These are open questions. How-
    ever, I have argued the importance of a number of goods: truth; politics;
    justice; aesthetics; inspiration and the spiritual. And I have also, albeit
    briefly, touched on the personal. This, then, is an open agenda. How to
    craft different goods, where, and in what balance, is for debate. Where, for
    instance, are we willing to decompose the truths of technoscience in favour
    of other goods? But there are no general answers. Specific questions and
    responses are needed.

6   *Imaginaries*. If we acknowledge that worlds and realities are multiple, then
    do we seek, nonetheless, to push towards singularity? Are there merits in
    the Euro-American insistence on the naturally definite? Or would it be a
    good to find ways of knowing and reality-making that allow the creation
    of many possible, more or less real, worlds? In this book I have argued in
    favour of allegory as a way of knowing the multiple and the ambivalent.
    I have also talked of 'gathering' as a way of avoiding discourses about
    coherence or consistency. But this is just a beginning. For instance, it also
    becomes important to find ways of crafting methods that do not seek to
    come to universal or general conclusions but do so specifically, location by
    location. So what would these look like? This is an open question, and
    there will be many answers.

7   *Materialities*. The question here is simple. Should we adopt a more generous
    and less exclusive approach to what can or should be made present in
    method? Its materialities? Should materials other than those that are
    currently privileged be recognised as presences that reflect and help to
    enact reality? Should we move beyond academic texts to texts in other
    modalities? And not just texts and figures, but bodies, devices, theatre,
    apprehensions, buildings? I have responded by saying yes to all these
    questions and have argued that the realities we know – and help to enact

– in academic texts, though important, are much too restricted. I have suggested that allegory is often likely to demand novel materialities. Once again, however, this is work to be done. There is need for a whole range of materially innovative methods.

8   *Indefiniteness*. I have argued that the dominant truth-related method assemblages tend to expect definite results and so enact definite realities. The question is: is this a good, or is it too restrictive? My response has been that it is too restrictive. Instead, I have argued that our methods should sometimes, perhaps often, manifest realities that are indefinite, and that as a part of this, it is important to appreciate that allegory, non-coherence, and the indefinite are not necessarily signs of methodological failure.

9   *Re-enchantment*. Euro-American method assemblages are dualist in effect, removing independent agency from the world of the real. The questions here are: should this dualist-inspired production of the real be weakened or abandoned? Are 'natural' realities possible agents? In this book I have argued that it is time to undo that dualism. Or, more precisely, I have argued that it is time to undo the Othering that underpins it, an Othering that conceals the enchanting complexities that generate the appearance of dualism. The flux and the resonances or patterns that can be made and detected in that flux are themselves a mode of enchantment. But somehow, in our common methods, we not only determine the location of agency but we also attempt comprehensive and systematic disenchantment.

Ending

These, then, are issues of *ontological methodology*. Their particular form reflects my own concerns and agendas. Enactment, multiplicity, fluidity, allegory, resonance, enchantment, these have been some of my keywords as I have explored what I have called method assemblage. But my object has been to provoke debate about methods rather than imposing a new orthodoxy. It is like this. If realities are enacted then many of the methodological certainties of the social and the natural sciences are undone and we need debate about what follows. Concern with the truth will not and should not go away. But the distinction between truths and other goods is at best pragmatic. All sorts of assemblages resonate to produce truths in one way or another. And our methods are implicated in other goods, political, aesthetic, spiritual, inspirational, or personally passionate (the list is not complete).

So what might one hope for method in a world where there are so many versions of the good? Again there will be no general 'best'. But I want to conclude by suggesting that it might be helpful to distinguish between what one might think of as 'procedural' and 'organisational' issues.

Procedural issues concern *how to conduct studies well*. About which goods to build into particular studies and in which forms. About how to reflect and enact particular commitments to (for instance) truth, or elegance, or politics, in an investigation. What is it to investigate a railway accident well? What are the approaches, the methods, that might be crafted to know about safety or pain or confusion? Procedural concerns, then, might look like a greatly broadened and at the same time much more modest version of our current methodological debates. They would be greatly broadened because they would reflect not only on how to make truths, but also on how to make other goods. Why did the trains collide? Yes. But what does it mean, 'why'? What realities are being made manifest or Othered in this or that mode of inquiry? Why do we make realities in this way or that? Is there a place for that that cannot be spoken? Which are the *goods* being made manifest or Othered? Which might we press? Or how might they be related? Indeed should they be related?

Debates of this kind would simultaneously be both broader and more modest than our current discussions of method. They would be more modest because they would arrive at *particular* conclusions in *particular* locations for particular studies. And there would be an allergy to general rules of methodological, political, aesthetic, or any other kind of hygiene. To any general constitutions. Not because there are not different goods or because it is not worthwhile going after them or linking them together, but because there is no longer any general way of moving effortlessly from place to place without attending to specificities. There *is* no general world and there *are* no general rules. Instead there are only specific and enacted overlaps between provisionally congealed realities that have to be crafted in a way that responds to and produces particular versions of the good that can only ever travel so far. The general, then, disappears, along with the universal. The idea of the universal transportability of universal knowledge was always a chimera.[135] But if the universal disappears then so too does the local – for the local is a subset of the general. Instead we are left with situated enactments and sets of partial connections, and it is to those that we owe our heterogeneous responsibilities.

Alongside such procedural questions there are also issues of *organisation*. For what I have been describing marks the beginning of the end of the modern constitutional settlement with its divisions of labour – divisions of labour that try to distinguish, as we have seen, between the truths of technoscience, the aesthetics of the arts, the rights and wrongs of politics or justice, the spiritualities of the religious, and the emotions and embodiments of the personal. There are, of course, very good reasons for making distinctions between these. The argument that truths are created more easily when they are detached from the political, may well be right. The division of labour also has the advantage that truths sometimes turn out to be politically subversive.

Thus to question the modernist constitution with its insistent division of labour is not to advocate collapse to some undifferentiated utopian social and technical order. The call is not to move towards a society without a division of labour. There is no perfect place, and surely we do not need a society in which every inquiry reflects a simultaneous commitment to truth, politics, beauty and all the rest of the possible goods. This would be the call for a totalitarianism run riot, and since out-thereness is lumpy and fractional, it makes little ontological let alone political sense. Matters are much more complex, and single recommendations no longer apply everywhere. There *is* no universal.

The problem, rather, is how to think well about the modes of relating between sites and specificities. These are not split off from one another by acts of God or cartographic men. Science and politics and aesthetics, these do not inhabit different domains. Instead they interweave. Their relations intersect and resonate together in unexpected ways. There are sets of partial connections and interferences. The issue, then, is about how to think and act these well – which is why I call it an organisational question. For it appears that the walls of the disciplines in the academy are very permeable, not only reflecting the ever-present requirement that truths should also be useful, but in the much wider and more creative sense that I have tried to condense in this book.

What does this mean in practice? The answer is that I do not know. But one thing is indeed clear. In the longer run it is no longer obvious that the disciplines and the research fields of science and social science are appropriate in their present form. It is no longer obvious that a division of labour is desirable, a division of labour that rests on the parcelling out of patches of truth to different specialists who are then divested of the need to practise other goods. After the subdivision of the universal we need quite other metaphors for imagining our worlds and our responsibilities to those worlds. Localities. Specificities. Enactments. Multiplicities. Fractionalities. Goods. Resonances. Gatherings. Forms of craftings. Processes of weaving. Spirals. Vortices. Indefinitenesses. Condensates. Dances. Imaginaries. Passions. Interferences. These are some of the metaphors for imagining method that I have sought to bring to life in this book. Metaphors for the stutter and the stop. Metaphors for quiet and more generous versions of method.

Glossary

Absence: the necessary Other to presence, which is enacted along with the latter, is constituted with it, and helps to constitute it. In method assemblage two forms of absence are distinguished. Manifest absence is that which is absent, but recognised as relevant to, or represented in, presence. Absence as Otherness is that which is absent because it is enacted by presence as irrelevant, impossible, or repressed. See also Otherness.

Actor-network theory: an approach to sociotechnical analysis that treats entities and materialities as enacted and relational effects, and explores the configuration and reconfiguration of those relations. Its relationality means that major ontological categories (for instance 'technology' and 'society', or 'human' and 'non-human') are treated as effects or outcomes, rather than as explanatory resources. Actor-network theory is widely used as a toolkit in sociotechnical analysis, though it might be better considered as a sensibility to materiality, relationality, and process. Whether it is a theory is doubtful. In the course of its development it has taken a wide range of different and sometimes inconsistent forms. It has at different times been criticised for its relative lack of interest in major social asymmetries such as gender, its refusal to base its explanations on generally accepted ontological categories, its tendency to a centred managerialism, the flattening character of its network metaphor, and its lack of concern with Otherness. The extent to which these complaints are appropriate to either early or contemporary work within the tradition is a matter of judgement.

Allegory: the art of meaning something other than, or in addition to, what is being said. The art of decoding meaning, reading between the literal lines to understand something else or more. The craft of making several things at once, what is described and what can also be read into that description. Ubiquitous, but often repressed into Otherness in contemporary standard understandings of representation.

Anteriority: out-thereness considered as prior to the process of knowing it. One of the assumptions made in standard versions of realism.

Condensation: crafted presence that may take a range of material forms.

Constructivism: the claim that scientific statements or truths are constructed in a way that to a large degree (in some versions totally) reflects the social circumstances of their production. Though there is some overlap, the programme of social constructivism is distinguished from the enactment approach of the method assemblage. Construction usually implies that objects start without fixed identities but that these converge and so gradually become stabilised as singular in the course of practice, negotiation and/or controversy. Enactment does not necessarily imply convergence to singularity, but takes difference and multiplicity to be chronic conditions.

Crafting: the enactment and condensation of presence in method assemblage. There is no implication that crafting is necessarily a human activity.

Critical realism: a contemporary and politically radical version of realism. Building on the realist suggestion that empirical and experimental investigation is unintelligible in the absence of an external world, and human capacity to intervene in that world and monitor the results of their action, it argues that the world is composed of objects, structures and causal or other powers, and that it is the job of the scholar to offer revisable theories or hypotheses about these. A distinction is made between the empirical (what appears in experience), the actual (actions that occur when powers or structures are activated), and the real (that which is there, those structures and powers, whether or not this is visible or activated). This means that empirical appearances, though important, may be misleading. It also means that the real may or may not be revealed by the actual, and there is no secure way of determining what is real. Distinguishing between the intransitive (roughly such objects of knowledge) and the transitive (the theories or terms used in knowledge), it notes that the transitive is socially located and variable, whereas the intransitive is not. No claims are made about the veracity or authority of the transitive domain, because theories or terms may be refuted and replaced by alternatives. In the terms proposed in this book, realism and critical realism are committed, at least in general, to the singularity, anteriority, independence, and probably to the definiteness of the real, as well as its primitive or originary version.

Cyborg: a trope from Donna Haraway's feminist material semiotics. This is a set of partial connections between two or more parts that cannot be reduced to one another but nonetheless relate to one another. Those parts may be material (between machine and human, or human and animal), political (as between different political or social identities and commitments), or they may exist in a tension between reality and fiction. The cyborg is a politically generative trope. It enacts possible novel realities by operating on and within material semiotic relations.

Deconstruction: see post-structuralism.

Deferral: an expression of the post-structuralist proposal that to make present is also, and at the same time, to make absent. Deferral is the removal and effacement of necessary absence into the future.

Definiteness: the assumption that out-thereness or absence is definite in form. One of the assumptions made in standard versions of realism.

Difference, problem of: the simultaneous existence of different objects that are said to be the same. This arises, as Annemarie Mol shows, because if objects are enacted in practices, and those practices are different, then so too are the objects that they produce, even if the practices in question are said to relate to, or be aspects of, the same object. Problems of co-ordination or separation then arise in the relations between the practices/objects.

Discourse: in its Foucauldian version, a set of relations of heterogeneous materiality, that recursively produces objects, subjects, knowledges, powers, distributions of power. Discourse is productive. At the same time it sets limits to what is possible or knowable.

Enactment: the claim that relations, and so realities and representations of realities (or more generally, absences and presences) are being endlessly or chronically brought into being in a continuing process of production and reproduction, and have no status, standing, or reality outside those processes. A near synonym for performance, the term is possibly preferable because performance has been widely used in ways that link it either to theatre, or more generally to human conduct.

Ends: see means and ends.

Enlightenment: a philosophically classical commitment to knowledge as the product of reason, empirical inquiry, and as a tool for social improvement. Historically, a period and a movement in eighteenth-century Europe.

Episteme: in Foucault's archaeology, a set of strategies laid down, permeating and producing the social body, which produce possibilities but also set limits to the conditions of possibility. See also discourse.

Excess: that which cannot be contained within narrative or linguistic discourse, but is probably also necessary to it. A version, or a way of talking about, Otherness.

Fallibilist method: an approach to method that both treats its theories, truth claims or propositions as refutable, and seeks to refute them on the grounds that in the longer run this is the best way to increase the power, scope, or veracity of knowledge. Associated with the work of Karl Popper, and now with realism and critical realism.

Feminist technoscience studies: a diverse body of empirical and theoretical work on the character of technology and science inspired by feminist theory and politics. Major themes or traditions of work include:

(a) So-called empiricist feminism which might seek to describe gender inequalities in science and technology.

(b) Epistemological critique, which explores the gendering built into scientific method and scientific findings which result from the social shaping of science.

(c) Standpoint epistemology, which argues that truth, or at least a workable version of knowledge, is most likely (or indeed only possible) from subordinate viewpoints, and perhaps particularly those of women or feminists.

(d) Material semiotics, which explores and seeks as a liberatory project, to interfere with the relations, simultaneously material and semiotic, that are enacted as partially connected patterns of practice, knowledge, subjectivity, objectivity and domination, by diffracting these in order to make a difference. Material semiotics privileges partial perspective, split vision and situated knowledge, arguing both that there is no escape from location and that identities, locations of knowledge, politics, and action are heterogeneous and irreducible rather than being coherent.

Flux: the sense that whatever is out there is not a structure with a discoverable shape, but is excessively filled with and made in heteromorphic currents, eddies, flows, vortices, unpredictable changes, storms, and with moments of lull and calm.

Fractionality: a metaphor for expressing the idea that objects, subjects and realities (and so their hinterlands) are more than one and less than many. The idea that hinterlands partially intersect with one another in complex ways. A way of avoiding two equally unsatisfactory alternatives: on the one hand the idea that multiplicity and difference imply ontological (and political) pluralism in which there are no interactions between multiples and realities proliferate without restraint, in a version of relativism; and on the other, the converse commitment to ontological singularity in which the world is taken to be singular and consistent.

Gathering: a metaphor like that of bundling in the broader definition of method assemblage. It connotes the process of bringing together, relating, picking, meeting, building up, or flowing together. It is used to find a way of talking about relations without locating these with respect to the normative logics implied in (in)coherence or (in)consistency.

Hinterland: a bundle of indefinitely extending and more or less routinised and costly literary and material relations that include statements about reality and the realities themselves; a hinterland includes inscription devices, and enacts a topography of reality possibilities, impossibilities, and probabilities. A concrete metaphor for absence and presence.

Idealism: see philosophical idealism.

Imaginary: a 'repertoire by which the world can be re-imagined, and in being re-imagined be re-made' (Verran).

Indefiniteness: see definiteness.

Independence: a commitment to the idea that whatever is out-there is usually independent of our actions and perceptions.

In-hereness: whatever is made present (for instance a representation or an allegory) that relates to and stands for whatever is made absent but depicted or connoted.

Inscription device: a system (often including though not reducible to a machine) for producing inscriptions or traces out of materials that take other forms. It may be understood as a particular modality for mediating out-thereness and in-hereness.

Interference: the pattern that derives from the intersection of two waveforms. In Haraway's material semiotics, a metaphor for the vision, necessarily split, that replaces representation or mirroring by recognising that it is situated and, indeed, split. At the same time action that makes a political difference. See also cyborg.

Manifest absence: see absence.

Material semiotics: see actor-network theory, feminist technoscience studies, cyborg and interference.

Materialism: see materiality.

Materiality: a way of thinking about the material in which this is treated as a continuously enacted relational effect. The implication is that materials do not exist in and of themselves but are endlessly generated and at least potentially reshaped. This is to be distinguished from materialism which, as the antonym of idealism, claims that what is real is material, and that the ideal is derived from material arrangements. Materiality makes no *a priori* distinction between the material and the ideal.

Means and ends: a hierarchical organising strategy that enacts and subordinates process or practice to the achievement of a valued goal. Therefore a mode in which most continuing processes of enactment are either Othered or are treated as techniques.

Mediation: the process of enacting relations between entities that are, as a part of that process, given form.

Metaphysics: in philosophy, untestable and often implicit assumptions which are enacted in and frame, experience or argument.

Method assemblage: generally, the process of crafting and enacting the necessary boundaries between presence, manifest absence and Otherness. Method assemblage is generative or performative, producing absence and presence. More specifically, it is the crafting or bundling of relations in three parts: (a) whatever is in-here or present (for instance a representation or an object); (b) whatever is absent but also manifest (that is, it can be seen, is described, is manifestly relevant to presence); and (c) whatever is absent but is Other because, while necessary to presence, it is also hidden, repressed or uninteresting. Presence may take the form of depictions (representational and/or allegorical) or objects. Manifest absence may take the form of a reality out-there that is represented, or the relevant context for an object. Method assemblage is distinguished from assemblage in the priority attached to the generation of presence. The definition by itself is symmetrical, telling us nothing about the form taken by presence, absence, or the relations between these. A further provisional definition of method assemblage is offered in Chapter 2. Here it is treated as the enactment of a bundle of ramifying relations that generate representations in-here and represented realities out-there. This is a special case of the more general definition above.

Modalities: conditions or contexts added to statements about reality that in one way or another tend to qualify them, sometimes undermining their authority.

Multiplicity: like difference, the simultaneous enactment of objects in different practices, when those objects that are said to be the same. Hence the claim that there are many realities rather than one. This arises because practices are endlessly variable and differ from one another. The additional claim that practices overlap in many and unpredictable ways, so there are always interferences between different realities. Multiplicity is inconsistent with singularity, but also with pluralism.

Object: a crafted version of condensed presence that takes the form of a process or entity deriving from and re-enacting an ordered form of absence. See method assemblage.

Ontological politics: if realities are enacted, then reality is not in principle fixed or singular, and truth is no longer the only ground for accepting or rejecting a representation. The implication is that there are various possible reasons, including the political, for enacting one kind of reality rather than another, and that these grounds can in some measure be debated. This is ontological politics.

Ontology: the branch of philosophy concerned with what there is, with what reality out-there is composed of.

Ostension: the process of defining a term by pointing to the object or event to which it refers.

Otherness: that which is neither present, nor recognisably or manifestly absent, but which is nevertheless created with, and creative of, presence. More strongly, that which is both necessary to presence, but necessarily pressed into absence or repressed. See also absence.

Out-thereness: the apprehension, common in Euro-American and many other cosmologies, that there is a reality outside or beyond ourselves. This may be specified and strengthened in a number of ways. See: primitive out-thereness; independence; singularity; and definiteness.

Performativity: the claim that words have effects on reality. More generally, the claim that enactments produce realities.

Philosophical idealism: a branch of philosophy which claims that what is real is non-material – for instance taking the form of the ideal or the spiritual – and that the ideal acts to produce the appearance of the material.

Pluralism: the idea that views or, more generally, realities, may co-exist in different locations without interfering with one another so long as appropriate ground rules can be put in place to regulate their relations and secure their independence. Hence a version of singularity (since ground rules would need to be shared by all). Therefore to be distinguished from multiplicity.

Post-structuralism: a middle and late twentieth-century philosophical movement which attacks what it takes to be a metaphysics of presence by arguing that attempts to bring everything to presence (for instance in the form of transparent representation) are flawed. This is because presence necessarily demands absence: the two are created or come into being together. One implication is that however complete representation may seem to be, it will reveal traces of Otherness, absence, or whatever is necessary to presence that has also been repressed. (Deconstruction is the analysis of texts and other presences to reveal traces of absence or Otherness.) A second implication is that the process of making present also produces that which is Other or absent. A third implication is that whatever is outside presence is unruly and excessive, perhaps to be sensed as flux. A fourth implication is that particular enacted versions of reality set limits to what they are able to know or create. Terms such as 'discourse', 'deferral' or 'episteme' point to such limits. Though the texts of post-structuralists are frequently taken to be abstract and philosophically demanding, many writers associated with or influenced by the approach (though they may resist the label) are also empirical or historical in a relatively straightforward way (for instance Foucault, Latour, Haraway, Mol).

Primitive out-thereness: the sense that there is a reality out there beyond ourselves. No particular claim is made about the character of that reality.

Realism: an approach to the philosophy of science that argues that empirical and experimental investigation is unintelligible in the absence of an external world, and human capacity to intervene in that world and monitor the results of their actions. See also critical realism.

Relativism: the idea that anything is as good as anything else, and there are no grounds for judging between them. This comes in at least three variants. Epistemological relativism says that the knowledge in your culture is just as good as the knowledge in my culture. There are no grounds for claiming that my account of out-thereness is any better than yours. Ethical relativism says that ethics are situated and local, and there are no grounds for claiming that my ethical standards are any better than yours. Political relativism takes the same form again: there are no reasons for preferring my politics over yours. We should live and let live. Relativism is closely related to pluralism, and is well understood as the other to singularity. It is to be distinguished from multiplicity, and the generation of fractionality in practices, where different realities, knowledges, ethics and politics are partially connected and interfere with one another.

Representation: a crafted version of condensed presence that depicts and re enacts manifest absence, while claiming or implying that its depictions are relatively direct expressions of manifest absence. See method assemblage.

Romanticism: in philosophy the idea that the world is so rich that the stories we might tell about it are irreducible either to one another, or (in some cases) to a single set of overall processes at all. The simultaneous claim that it is important not to lose that richness. Historically, a reaction to the rationalism of the Enlightenment.

Singularity: the idea that there are definite, limited, and therefore single, sets of processes in the world, that the world is a single thing.

Stop: a version of deconstruction, in which a smooth narrative that has been brought to presence displays a break or an interruption that opens up the uncertainties of Otherness.

Symbolic interactionism: a predominantly American tradition in sociology based in the analysis of practice, and treating knowledges and identities as being produced, and irremovable from, particular practices. Strongly influenced by philosophical romanticism, it is relatively sceptical about Enlightenment or classical claims that knowledge can be formalised and transmitted apart from practices and cultures.

Symmetry: the principle that the same kind of explanation or account should be given for all the phenomena to be explained. In the context of science this means that the truth or falsity of scientific ideas should be ignored, and all should be explained in the same general terms. In the present book the principle is applied to method. Method assemblage is a way of thinking about all methods in the same terms, whether or not these fit normative rules about social science method.

Universalism: the idea that true knowledge derives from universal criteria that can and should be applied in all relevant contexts. Hence the idea that true knowledge does not vary between context.

View from nowhere: a way of talking about the idea that we can step outside and so obtain an overview of the world that is detached from any particular location or practice.

Notes

1   The literatures are extensive, and I cannot possibly survey them here. Indicative citations would include Clifford and Marcus (1986) and Haraway (1989) on poetics, Ashmore (1989) and Latour (1996) on reflexive methods and multiple narratives, Haraway (1997) and Rose (2001) on visual methodologies, Butler (1993) and Thrift (2000) on textual and embodied narrative, and Clifford (1997), Hine (2000), Thrift (2000) and Urry (2000) on geographically distributed methods.

2   See Doll and Hill (1950).

3   For an introduction see Nettleton (1995, 160ff).

4   See Klinenberg (2002).

5   For some of the possible complexities, worked out for the example of the UK cervical screening programme (It looks like a success, but is it? If it is a success then how is it so?) see Singleton (1998). I discuss this further in Chapter 5.

6   I will use this term as an index of a more or less hegemonic set of claims about method, notwithstanding the divergences in practice. For an account of its considerable difficulties see Ingold (2000). There are many studies that explore the construction and social correlates of social (and natural) science. I consider the division of labour between truth and politics briefly below (see Shapin and Schaffer (1985) and Haraway (1997)). See also the work by Theodore Porter and Ian Hacking on the contingency of the relations between quantification and scientific (including social scientific) inquiry (Hacking 1990; Porter 1995).

7   The power but also the limits of auditing are considered in Michael Power (1997).

8   This formulation ignores important differences within the STS literatures. Some of these are considered in later chapters.

9   For recent exemplary cases see, for instance, Campbell (1987) in *verstehende* sociology, Becker (1982) in symbolic interactionism, Said (1991) in postcolonialism, Latour (1996; 1998) in (so-called) actor-network theory, and Haraway (1991b) in feminist technoscience studies.

10 Symbolic interaction offers us an exemplary case of an approach to method largely romantic in inspiration which then cut its cloth to fit the much more definite and determinate picture of the world imagined by post-World War II sociology in the United States. Consider, for instance, the assumptions built into the method of grounded theory. For an admirable historical and philosophical overview see Rock (1979).

11 It is systematised in this mode in particular by Karl Popper. See Popper (1959).

In a more contemporary context realism and critical realism present themselves as fallibilist methods. See, for instance, the description in Benton and Craib (2001).

12   Latour says similar things about theory when this is imagined as something that can be rapidly displaced with ease. Not so, he says. In practice it takes a huge amount of work. See Latour (1988).

13   The slogan is similar to Paul Feyerabend's much misunderstood philosophy of science. His commitment to methodological anarchism derives from his assumption that a proliferation of methods would generate the best and most rigorous science. See Feyerabend (1975), and for its translation into social science, Phillips (1973).

14   I draw the notion of entanglement and disentanglement from Michel Callon. See Callon (1998a).

15   See in particular, Knorr Cetina (1981) and Lynch (1985).

16   This account draws on Alpers (1989), Bryson (1983), Law and Benschop (1997) and Rotman (1987).

17   The use of mirrors and optics of all kinds was almost certainly crucial from the fifteenth to the nineteenth centuries in the development of European fine art. See Hockney (2001).

18   See, for instance, the illustration from Jan Vredeman de Vries at http://www. kb.nl/kb/100hoogte/hh-im/hh046.html (from the web page of the Koninklijke Bibliotheek *Nationale bibliotheek van Nederland.*)

19   As, for instance, in the Annunciation by the Master of the Barberini Panels in the National Gallery of Art in Washington DC. See http://www.nga.gov/cgi-bin/pimage?362+0+0+gg4.

20   Raphael. *Marriage of the Virgin.* 1504. Oil on panel. Pinacoteca di Brera, Milan, Italy.

21   For instance, 'idealism' is an ontology that says that in the first instance there is nothing material. Everything, including the material, is produced by the spirit, the mind, or the process of knowing. 'Materialism' is an ontology that says, contrariwise, that everything is material. 'Spirit' or 'mind' are expressions of the material. The latter is well known in social science, in part through the Marxist tradition. Notoriously, in his historical materialism Marx stood (the idealist) Hegelian dialectic on its head.

22   I write 'usually' because we also appreciate that sometimes our actions affect parts of that external reality – and this is especially the case for social reality. Philosophical realists distinguish, for instance, between the transitive and the intransitive. For a convenient survey see Sayer (2000).

23   In Chapter 7 I will consider a cosmology, that of Australian Aborigines, where this appears to be the case.

24   Sometimes, indeed, claims that were previously unqualified may be 'modalised' and start to lose their authority.

25   An attractive version of this argument presented in a slightly different idiom is to be found in Collins (1975). I return to this in a later chapter.

26   This is also the case for instruments which work well in one location, but fail to do so in another. For a nice case see Collins (1974).

27   The sociologists of science sometimes call this 'black-boxing'.

28   See Stengers (1997).

29   See, for instance, Foucault (1970; 1972; 1979).

30 See, for instance, Rose (1999).

31 The point also applies to Latour and Woolgar's own claims. They too are caught up in (and helping to produce) an obdurate hinterland – which includes the Euro-American common-sense experience that out-thereness is obdurate, anterior and all the rest. Accordingly, their position is internally consistent.

32 The development of quantitative data collection and related tests of significance are the subject of a considerable literature. See, for instance, Hacking (1990) and Porter (1995). Timekeeping is the subject of a large literature: see the classic Thompson (1967), and for a convenient summary Thrift (1996).

33 See Osborne and Rose (1999) and Law and Urry (2004).

34 This is a mild way of putting what can be a much stronger point. Feminist technoscience studies have in particular pressed for the enactment of interfering research programmes with appropriate theoretical and methodological tools. Donna Haraway's work on a non-militaristic, non-sexist, non-racist cyborg is particularly well known. See Haraway (1991a).

35 The importance of symmetry was first emphasised in these terms by Bloor (1976), though it is implicit in the work of such historians as Kuhn. I return to the topic more fully at the end of Chapter 5.

36 See, for instance, Deleuze and Guattari (1988) and Deleuze and Parnet (1987).

37 These include (in translation) the following verbs: to fit up, adapt, adjust, reconcile, bring into accord, settle, dispose, arrange, combine, unite, compose, constitute, form, co-ordinate, organise, lay out, prepare, distribute, deal out, chain, tie down, link up, connect, order, array, settle, place, put, set, lay, put out, join together, gather, assemble, muster, collect, bring together, and/or unite. The small French–English dictionary is the *Concise Oxford French Dictionary* (Chevalley and Chevalley 1963), and the larger French dictionary is the large Robert (1974). I am grateful to Michel Callon for discussion of the difficulties of the term 'assemblage' in English.

38 Cooper (1998, 111); and the translator's introduction to Deleuze and Parnet (1987, xiii).

39 Perhaps it sounds as if it has to do with the action of assembling – for instance as in school, army or prison musters, or perhaps the process of gathering together things on a list, as if one were packing before travelling.

40 Libraries have been written about this, and we need only the sketchiest account here. Good places to start in a review of these debates include: Kuhn (1970), Lakatos and Musgrave (1970) and Barnes (1982).

41 This is a crucial Kuhnian lesson – though it comes from other authors and other literatures too. See, for instance, Polanyi (1958) and Ravetz (1973).

42 This has several radical implications. One is that since there are scientific revolutions, discontinuities in the history of science, it is not so very easy to show that science progresses. Perhaps it merely changes. Indeed Kuhn got into a lot of trouble with his critics because he claimed that since scientific revolutions are discontinuities this means that science itself advances discontinuously. Thus most previous accounts of scientific change assumed that in general and over time science increased its predictive power, the scope of its theories, and its empirical base. This argument was made in a variety of different ways, but usually assumed that science created generalisations of increasing power and parsimony, and/or falsified those that turned out to be empirically inadequate. But this (usually) implies some kind of empirical yardstick for measuring the

scope of scientific theories. Sure, scientists – or whole groups of scientists – might get hold of the wrong end of the stick, and fool themselves into thinking they'd discovered phenomena that weren't actually there. But overall, and in the long run, it was assumed that good observation would out, so long as the process of inquiry was disentangled from the malign effects of political and economic interference. On falsification, see Popper (1959).

43   Thus it turns out that if patients start regular walking under the appropriate supervision of physiotherapists, many report that the onset of pain is increasingly delayed, and sometimes it is not necessary to operate at all. See Law and Mol (2002).

44   Interestingly, when walking therapy works (which it usually does only with the support and discipline of physiotherapy) it does not appear to reduce stenoses. So why does it work? Perhaps it opens up alternative 'collateral' vessels which bypass the diseased arteries. Perhaps it alters the biochemistry of the blood. No one actually knows.

45   The Salk scientists do too, in practice. As we have seen, they live in uncertainty. But unlike the medical professionals, they set themselves the convergent goal of determining a single reality.

46   The approach is common in the sociology of scientific knowledge (SSK). See, for instance, Collins (1975; 1981a).

47   He brilliantly developed this through a series of studies, starting with Goffman (1971).

48   As this suggests, the turn to the performativity of enactment has been a powerful if not dominant force in a number of theoretical traditions for several decades. It would be possible to write a genealogy of this opening as it has struggled first to imagine (or enact?), and then to come to terms with, the epistemological, ontological and theoretical implications of the idea that the real is enacted in practices, rather than being reflected through them, as it is in perspectivalism. Louis Althusser (1971) works uncomfortably in the space defined by these two possibilities. Michel Foucault (1979; 1981) is much clearer about the performativity of discourse, as are STS writers such as Latour and Woolgar, and such feminist theorists as Judith Butler (1993) and Donna Haraway (1997). Any serious attempt to imagine the performativity of enactment also has to handle the related question of materiality or 'the real' and its relations with discourses or other linguistic expressions. In Foucault, discourse extends into and is carried through certain kinds of materials – an opening explored more fully for the case of embodiment by Butler.

49   For a more extended discussion see Law (2002a).

50   For further discussion of non-coherent hierarchies or (more generally) intransitive relations see Law (2000).

51   This argument is developed at greater length in Law (2002a).

52   For critical radical commentary on such identity politics see, for instance, Haraway (1991a) and Harvey (1993).

53   In his book *We Have Never Been Modern*, Latour (1993) argues that what is sometimes called modernity is productive precisely because it insists on purity. It insists, for instance, that things have single and definite shapes, or that natural realities are clear and quite distinct from those of the social. His argument is not that that modernity actually achieves this purity. Rather it is that by imagining reality to be pure it allows the fecund production of

impurities – swarms of heterogeneous multiplicities. Latour argues that we have never been modern. We just think that we are.

54 See Lacqueur (1990).

55 The argument is developed in Mol (2002), Hirschauer (1998) and Hirschauer and Mol (1995). Hirschauer's work attends to the issues for sex/gender as these arise for transsexuals, a difficult context far from the abstractions of theory. It is also developed, in a slightly different idiom, in Cussins (1998a; 1998b; 1998c).

56 See Haraway (1991b, 194–195) and (1997).

57 For commentary see Moser (2000).

58 For details of those publications see Law and Singleton (2003; forthcoming).

59 In order to preserve anonymity, all proper names and locations in what follows are pseudonyms, other than those of national organisations.

60 Dr Warrington, a consultant gastro-enterologist, was interviewed on 19 March 1999. Quotations are reconstructed from notes.

61 Dr Warrington, 19 March 1999.

62 Dr Warrington, 19 March 1999.

63 The quote is not from Sister Fraser but her senior colleague, a Nursing Officer. Interview on 10 December 1998.

64 Sister Fraser was interviewed on 10 March 1999. Quotations are reconstructed from notes.

65 Sister Hart was interviewed on 3 March 1999. Quotations are reconstructed from notes.

66 Dr Bowland was interviewed on 11 June1999. Quotations are reconstructed from notes.

67 For a fine study of the distribution of responsibility between individuals and social arrangements, and the individualisation of responsibility, see the related case of the 'problem' of drink driving by Gusfield (1981).

68 This is an instance where the purification described by Latour and discussed in the last chapter seemed to impede the proliferation of impure forms.

69 Excerpt from letter sent to the initiating hospital consultant dated 7 October 1998.

70 From interview notes with the staff at Castle Street Centre, Sandside, 10 June 1999.

71 Interview, 17 June 1999.

72 Related metaphors for fluid objects have been developed in a number of other contexts. See, for instance, Mol and Law (1994), Law and Mol (2001) and Law (2002c).

73 And there are other analyses that have a similar shape. See, for instance, my own account of technological decision making briefly discussed above, and more fully in Chapter 7 of Law (2002a).

74 Materiality, not materialism, since the argument is not reductionist.

75 See, for instance, Butler (1993).

76 Cussins (1998a; 1998b; 1998c); Moser (2000; 2003); Moser and Law (1998; 1999; 2003); Moreira (2000; 2001a; 2001b).

77 This relational metaphysics is laid out systematically in Latour (1988) and (1998). For further commentary see Law and Mol (1995).

78 Routine is what gets hidden because whatever is in front of it (presence and manifest absence) includes it and hides it. In STS this is sometimes known as black-boxing (Rip 1986). Examples such as the workings of a personal

computer, hidden while all goes well, explain why the black-box metaphor is appealing. Insignificance is not so different but is less discrete. Repression indexes a lively and important tradition running from Freud through versions of post-structuralism (for instance in the writing of Lacan and Lyotard) to a range of radical interventions in cultural studies that have often explored how subordinates (for instance blacks or women) are Othered to produce versions of white male superordination. See, for instance, Hall (1992), Said (1991), and Haraway (1989).

79   For representation, on the face of it to talk about something other than what one is talking about is at best roundabout, perhaps a metaphorical flourish, and at worst it is simply misleading.

80   These skills, to be sure, work the other way round. The powerful treat the representations of the less powerful allegorically too, doubting, cross-examining, checking and auditing. Trust is in short supply in both directions. On the self-defeating character of the audit process, which can be understood as a futile attempt by the powerful to convert allegory into representation, see Power (1997).

81   Though arguably sociology has been important in the enactment of the social. For hints to this effect see Porter (1995), Osborne and Rose (1999) and Law and Urry (2004).

82   For discussion of economics see Callon (1998b).

83   Those not caught up directly in the relations enacted in these claims do not necessarily take those claims at face value. Remember, however, that economic realities are not simply statements but are also relations that extend into practices and materials that ramify off in all directions. 'Belief' is not usually what is at stake.

84   Their worry is no doubt compounded by the suspicion that lack of public support affects the depth to which public bodies are willing to reach into their pockets to fund scientific research. Thus, surely, is one of the explanations for the so-called 'science wars' controversies in which social scientists have been accused of undermining the epistemological foundations of natural science.

85   The Minister in question, John Selwyn Gummer, was responding to fears about new-form Creutzfeldt-Jakob Disease at the beginning of the BSE scare in the UK, in a dramatic televisual attempt to persuade the nation that beef was indeed safe. This took place on 6 May 1990 (see http://news.bbc.co.uk/1/hi/uk/369625.stm). It is appropriate to note that his politics are now much greener than they were at that time.

86   These examples together with the larger argument about the public under-standing of science are drawn from Brian Wynne's work. See, for instance, Wynne (1996).

87   See Singleton and Michael (1993), and Singleton (1996; 1998).

88   This is an argument that she has developed in work on the UK campaign to reduce death from sudden infant death syndrome, or SIDS. Official statistics suggest that the campaign has been remarkably successful, following a 'back to sleep' campaign to persuade mothers to place their infants on their backs before they go to sleep. Singleton's data reveal that this injunction is interpreted and enacted in many different ways in practice.

89   An argument like this is developed by Frederic Jameson in his analysis of the 'post-modern' architectures of San Francisco, which, he argues, may be

understood as tools for what he calls 'cognitive mapping' that are appropriate to the non-coherent but global realities of capitalism. See Jameson (1991). For further discussion of 'knowing in tension', see Law (1998).

90 Ladbroke Grove Rail Inquiry, Transcript of Proceedings, morning of 10 May 2000, page 5. This was available at http://www.lgri.org.uk/10mayam.htm, now saved at http://www.archive.com.

91 The details are drawn from the following documents: Second Health and Safety Executive Interim Report, 'Train Accident at Ladbroke Grove Junction, 5 October 1999', 3 November 1999, http://www.hse.gov.uk/railway/paddrail/interim2.htm; Third Health and Safety Executive Interim Report, 'Train Accident at Ladbroke Grove Junction, 5 October 1999', 14 April 2000, http://www.hse.gov.uk/railway/paddrail/interim3.htm; Ladbroke Grove Rail Inquiry, Transcript of Proceedings, morning of 10 May 2000, formerly at http://www.lgri.org.uk/10mayam.htm (now saved at http://www.archive.com).

92 Many of the details including transcripts of the inquiry were available at http://www.lgri.org.uk. Regrettably, this website was closed in the summer of 2002, though most of the pages can be retrieved from the excellent facility at http://www.archive.com. See also Cullen (2001).

93 Much of the debate and cross-examination in the inquiry concerned the relative significance or plausibility of different possible causes. A straightforwardly allegorical reading of many of these interventions is irresistible. The protagonists were trying to ensure, as plausibly as they possibly could, that important contributory causes did not end up in their own backyard.

94 This is from Cullen (2001, 7) and is the terms of reference for Part 1 of that Inquiry. 'HSE' is the acronym for the Health and Safety Executive.

95 An issue about which I have learned much in discussion with Ingunn Moser who writes on disability and interferes in disability politics. See Moser and Law (1998; 1999); Moser (2000; 2003).

96 Thus Haraway's account of the cyborg is similarly allegorical, as is the concern with absence, Otherness, and intertextuality witnessed in the heritage of Foucault and Lacan, cultural and postcolonial studies and parts of feminist theory.

97 See Chapter 6.

98 Barnes distinguishes between a legitimate interest in the prediction and control of nature, and an illegitimate and concealed interest in social control and rationalisation. Science is generally, he says, and preferably, under the direction of the former. However, even the latter may produce cultural forms that are relevant to natural prediction and control. The origins of knowledge tell us nothing about its utility and validity. For this reason he does not distinguish between 'ideology' and 'knowledge', but talks instead of 'ideological determination'. See Barnes (1977).

99 The sociologists of scientific knowledge were here following a line of argument that has its hinterland in both social anthropology and the *verstehende* tradition in sociology.

100 This ethnography is reported more fully in Law (1994).

101 This argument has been elaborated into a much larger metaphysics in Lawson (2001), which, however, does not entertain the divergent possibilities of difference, multiplicity and fractionality.

102 For references see Pinch (1980; 1981; 1985).

103 See, for instance, the work of Knorr Cetina (1999), Pickering (1995) and Traweek (1988).

104 There is a tradition in the philosophy of science that formalises this. See Hesse (1963; 1974). The empirical studies in the sociology of scientific knowledge also show that what counts as 'right' and 'wrong' is often, perhaps always, negotiable – though, as Latour and Woolgar suggest, this may become so expensive as to be impossible. Collins's work is related to that of Latour and Woolgar, but there are also differences. Collins is particularly keen to show that descriptions of reality are located and grounded in cultures or forms of life, and is happy to describe himself as a relativist, a claim carefully avoided by Latour and Woolgar. The differences are debated in Collins and Yearley (1992) and Callon and Latour (1992).

105 This is a distinction that crops up in different but somewhat related ways in a range of different literatures. It resonates, for instance, with the distinction between classical and romantic thought described by Alvin Gouldner (1973). A similar theme is explored by Karl Mannheim in his essay on conservative thought (1953). Rather differently, Mary Douglas's anthropology distinguishes between more bureaucratic or ritualised settings, and those that are more entrepreneurial. See Douglas (1982). The present book, as I suggested in an earlier chapter, locates itself in a similar divide.

106 The study is reported more fully in Law (1994).

107 The numberings for *Christian Faith and Practice* refer to paragraphs, not pages.

108 See Pickering (1995).

109 See Shapin and Schaffer (1985), and Shapin (1994), together with Shapin (1989).

110 For further commentary on this see Haraway (1997).

111 See Alpers (1989), Haraway (1991b).

112 Additional references come from Robert Layton (1989).

113 See Kerle (1995, 136); the details are discussed more fully in Ayre (2002).

114 See Chatwin (1998).

115 Verran develops her argument so: 'the beginning of a *galtha* workshop emphasizes the multiplicity of practitioner groups and their differing contributions to the necessarily messy reality of a place, the opening scenario of a science practical hides differences between the many and varied practitioner groups that constitute the environmental sciences, invoking instead a virtual, singular place. These different assessments then switch when we proceed to what is actually done during the workshop. While for Yolngu it is important that multiple possible "doings" be channeled into one communal act of place, scientists need to perform, report, and make known a multiplicity of actual doings. I am suggesting, then, that the normal ontologies of these two knowledge traditions advance different ways of managing the multiplicity/ singularity tension that comes with doing any ontology of place' (Verran 2002, 165).

116 Aboriginal people originally congregated in stations as a means of living acceptably in or close to their own country. In the 1970s a law was passed which required White station owners to pay 'award wages' (minimum wages) to Aborigines for their work. This led the station owners to turn Aborigines off their lands and into mission settlements.

117 The implicit reference is to Latour (1993).

118 For an entertaining and partially fictional essay which explores this (and much else) see Julian Barnes (1990).

119 For further discussion of the 'imaginary' see Verran (2001; forthcoming).

120 For details see, for instance, Verran (1998) and Sharp (1996).

121 See Margaret Ayre's remarkable study (2002) of nature conservation and management in East Arnhemland. And David Turnbull's account (forthcoming).

122 See, for instance, Baskhar (1979), and for recent accounts in the context of social science, Sayer (2000) and Benton and Craib (2001).

123 Realists refer to this as the 'transitive' dimension of inquiry, in contrast with 'intransitive' natural phenomena.

124 For an account of the difficulties of tightly integrated systems see Charles Perrow's exemplary text, *Normal Accidents* (1984). For further discussion of non-coherent coherence see Singleton and Michael (1993), Singleton (1998) and Law (2002a).

125 It is the emphasis on presence that distinguishes *method* assemblage.

126 It may be that this repressed multiplicity is necessary to achieve the appearance of singularity, though under certain circumstances the contrary argument can also be made.

127 As we have seen, the argument is developed by Latour and Woolgar. But see, also, Latour (1990).

128 David Turnbull's exemplary work on cartography deserves careful study. See Turnbull (1993; 1996; 2000).

129 Here I am commenting on academic or other forms of writing that seek to describe realities. As is obvious, the argument does not necessarily apply in this form to non-referential forms of writing such as novels or poetry.

130 We have encountered it, for instance, in the writing of Donna Haraway. See Haraway (1991a; 1991b; 1997; 2003).

131 There are interesting accounts (of particular versions of aesthetics) in the physics described by Traweek. See Traweek (1988; 1999), and different aesthetic styles are implied in Turkle's work on computer use. See Turkle (1996).

132 Donna Haraway's recent work on people and dogs as companion species, though written in a very different idiom, makes an argument that is connected to this. See Haraway (2003).

133 This is the last line of her 'Cyborg Manifesto'. See Haraway (1991a, 181). Her argument is (necessarily) situated, in part by her erotic and political commitment to a refigured version of science in a context where it was easy to see science as inhumane and fundamentally flawed. For a further part of the relevant feminist political and spiritual context see Starhawk (1989).

134 It is also the case that symmetry is always a moving target. Thus the argument from symmetry assumes that everything can be made manifest. But Othering is a limitless domain. Only particular assumptions can be made manifest. The issue, then, is one of openness or attitude to the hidden realities of Othering, rather than enumerating a complete list of repressed asymmetries.

135 The argument is developed by such writers as Latour. See, for instance, Latour (1987).

References

Addelson, Kathy (1994), *Moral Passages: Towards a Collectivist Moral Theory*, New York and London: Routledge.

Alpers, Svetlana (1989), *The Art of Describing: Dutch Art in the Seventeenth Century*, London: Penguin.

Althusser, Louis (1971), 'Ideology and Ideological State Apparatuses (Notes towards an investigation)', pp. 121–173 in Louis Althusser (ed.), *Lenin and Philosophy and Other Essays*, London: New Left Books.

Appelbaum, David (1995), *The Stop*, Albany, NY: SUNY Press.

Ascherson, Neal (2002), 'Hitler's Teeth: Review of Berlin: The Downfall, 1945 by Antony Beever', *London Review of Books*, 28 November 2002, 15–16.

Ashmore, Malcolm (1989), *The Reflexive Thesis*, Chicago: Chicago University Press.

Ashmore, Malcolm, Michael J. Mulkay and Trevor J. Pinch (1989), *Health and Efficiency: A Sociology of Health Care Economics*, Milton Keynes: Open University Press.

Ayre, Margaret (2002), 'Yolngu Places and People: Taking Aboriginal Understandings Seriously in Land and Sea Management', PhD, University of Melbourne.

Bardon, Geoffrey, and Tim Leura Tjapaltjarri (n.d.), 'The Great Painting, Napperby Death Spirit Dreaming', pp. 46–47 in Judith Ryan (ed.), *Mythscapes: Aboriginal Art of the Desert from the National Gallery of Victoria*, Melbourne: National Gallery of Victoria.

Barnes, Barry (1977), *Interests and the Growth of Knowledge*, London: Routledge & Kegan Paul.

Barnes, Barry (1982), *T.S.Kuhn and Social Science*, London: Macmillan.

Barnes, Julian (1990), *A History of the World in 10½ Chapters*, London: Picador.

Baskhar, Roy (1979), *A Realist Theory of Science*, Hemel Hempstead: Harvester Wheatsheaf.

Bauman, Zygmunt (1989), *Modernity and the Holocaust*, Cambridge: Polity Press.

Becker, Howard S. (1982), *Art Worlds*, Berkeley: University of California Press.

Benton, Ted, and Ian Craib (2001), *The Philosophy of Social Science*, Basingstoke: Palgrave.

Bloor, David (1976), *Knowledge and Social Imagery*, London: Routledge & Kegan Paul.

Bryson, Norman (1983), *Vision and Painting: The Logic of the Gaze, Language, Discourse, Society*, Basingstoke: Macmillan.

Butler, Judith (1993), *Bodies that Matter: On the Discursive Limits of 'Sex'*, New York and London: Routledge.

Callon, Michel (1986), 'Some Elements of a Sociology of Translation: Domestication of the Scallops and the Fishermen of Saint Brieuc Bay', pp. 196–233 in John Law (ed.), *Power, Action and Belief: A New Sociology of Knowledge?* Sociological Review Monograph, 32, London: Routledge & Kegan Paul.

Callon, Michel (1998a), 'An Essay on Framing and Overflowing: Economic Externalities Revisited by Sociology', pp. 244–269 in Michel Callon (ed.), *The Laws of the Markets*, Oxford and Keele: Blackwell and the Sociological Review.

Callon, Michel (1998b), 'Introduction: The Embeddedness of Economic Markets in Economics', pp. 1–57 in Michel Callon (ed.), *The Laws of the Markets*, Oxford and Keele: Blackwell and the Sociological Review.

Callon, Michel, and Bruno Latour (1992), 'Don't Throw the Baby Out with the Bath School! A Reply to Collins and Yearley', pp. 343–368 in Andrew Pickering (ed.), *Science as Practice and Culture*, Chicago: Chicago University Press.

Campbell, Colin (1987), *The Romantic Ethic and the Spirit of Modern Consumerism*, Oxford: Blackwell.

Chatwin, Bruce (1998), *The Songlines*, London: Vintage Classic.

Chevalley, Abel, and Marguerite Chevalley (eds) (1963), *The Concise Oxford French Dictionary*, Oxford: Clarendon Press.

Clifford, James (1997), *Routes: Travel and Translation in the Late Twentieth Century*, Cambridge, MA and London: Harvard University Press.

Clifford, James, and George E. Marcus (eds) (1986), *Writing Culture: The Poetics and Politics of Ethnography*, Berkeley, Los Angeles and London: University of California Press.

Collins, H.M. (1974), 'The TEA Set: Tacit Knowledge and Scientific Networks', *Science Studies*, 4: 165–185.

Collins, H.M. (1975), 'The Seven Sexes: A Study in the Sociology of a Phenomenon, or the Replication of Experiments in Physics', *Sociology*, 9: 205–224.

Collins, H.M. (1981a), 'The Place of the "Core Set" in Modern Science: Social Contingency with Methodological Propriety in Science', *History of Science*, 19: 6–19.

Collins, H.M. (1981b), 'Son of Seven Sexes: The Social Destruction of a Physical Phenomenon', *Social Studies of Science*, 11: 33–62.

Collins, H.M., and Steven Yearley (1992), 'Epistemological Chicken', pp. 301–326 in Andrew Pickering (ed.), *Science as Practice and Culture*, Chicago: Chicago University Press.

Cooper, Robert (1998), 'Assemblage Notes', pp. 108–129 in Robert C.H. Chia (ed.), *Organized Worlds: Explorations in Technology and Organization with Robert Cooper*, London and New York: Routledge.

Crossman, Richard (1975), *Diary of a Cabinet Minister, Vol. 1, Minister of Housing, 1964–1966*, London: Hamish Hamilton and Jonathan Cape.

Cullen, Rt. Hon. Lord (2001), *The Ladbroke Grove Rail Inquiry, Part 1*, Norwich: HSE Books, Her Majesty's Stationery Office.

Cussins, Charis M. (1998a), 'Ontological Choreography: Agency for Women Patients in an Infertility Clinic', pp. 166–201 in Marc Berg and Annemarie Mol (eds), *Differences in Medicine: Unravelling Practices, Techniques and Bodies*, Durham, NC and London: Duke University Press.

Cussins, Charis M. (1998b), 'Producing Reproduction: Techniques of Normalization and Naturalisation in Infertility Clinics', pp. 66–101 in Sarah Franklin and Helen

Ragone (eds), *Reproducing Reproduction: Kinship, Power and Technological Innovation*, Philadelphia: University of Pennsylvania Press.

Cussins, Charis M. (1998c), '"Quit Sniveling, Cryo-Baby. We'll Work Out Which One's Your Mama!"', pp. 40–66 in Robbie Davis-Floyd and Joseph Dumit (eds), *Cyborg Babies: From Techno-Sex to Techno-Tots*, New York and London: Routledge.

Daston, Lorraine (1999), 'Objectivity and the Escape from Perspective', pp. 110–123 in Mario Biagioli (ed.), *The Science Studies Reader*, New York and London: Routledge.

de Laet, Marianne, and Annemarie Mol (2000), 'The Zimbabwe Bush Pump: Mechanics of a Fluid Technology', *Social Studies of Science*, 30: 225–263.

Deleuze, Gilles, and Félix Guattari (1988), *A Thousand Plateaus: Capitalism and Schizophrenia*, London: Athlone.

Deleuze, Gilles, and Claire Parnet (1987), *Dialogues*, London: Athlone.

Derrida, Jacques (1982), 'Différance', pp. 1–27 in *Margins of Philosophy*, Hemel Hempstead: Harvester Wheatsheaf.

Doll, Richard, and A. Bradford Hill (1950), 'Smoking and Carcinoma of the Lung', *British Medical Journal*, 2: 739–748.

Douglas, Mary (1982), 'Cultural Bias', pp. 183–254 in Mary Douglas (ed.), *In the Active Voice*, London: Routledge & Kegan Paul.

Feyerabend, Paul K. (1975), *Against Method: Outline of an Anarchistic Theory of Knowledge*, London: New Left Books.

Foucault, Michel (1970), *The Order of Things: An Archaeology of the Human Sciences*, London: Tavistock.

Foucault, Michel (1972), *The Archaeology of Knowledge*, London: Tavistock.

Foucault, Michel (1979), *Discipline and Punish: The Birth of the Prison*, Harmondsworth: Penguin.

Foucault, Michel (1981), *The History of Sexuality, Volume 1: An Introduction*, Harmondsworth: Penguin.

Goffman, Erving (1971), *The Presentation of Self in Everyday Life*, Harmondsworth: Penguin.

Gouldner, Alvin (1973), 'Romanticism and Classicism: Deep Structures in Social Science', pp. 323–366 in Alvin Gouldner (ed.), *For Sociology*, London: Allen Lane.

Gusfield, Joseph R. (1981), *The Culture of Public Problems: Drinking-Driving and the Symbolic Order*, Chicago: University of Chicago Press.

Hacking, Ian (1990), *The Taming of Chance*, Cambridge: Cambridge University Press.

Hacking, Ian (1992), 'The Self-Vindication of the Laboratory Sciences', pp. 29–64 in Andrew Pickering (ed.), *Science as Practice and Culture*, Chicago and London: Chicago University Press.

Hall, Stuart (1992), 'The West and the Rest: Discourse and Power', pp. 275–331 in Stuart Hall and Bram Gieben (eds), *Formations of Modernity*, Cambridge: Polity and Open University Press.

Haraway, Donna J. (1989), *Primate Visions: Gender, Race and Nature in the World of Modern Science*, London: Routledge.

Haraway, Donna J. (1991a), 'A Cyborg Manifesto: Science, Technology and Socialist Feminism in the Late Twentieth Century', pp. 149–181 in Donna Haraway (ed.), *Simians, Cyborgs and Women: The Reinvention of Nature*, London: Free Association Books.

Haraway, Donna J. (1991b), 'Situated Knowledges: the Science Question in Feminism and the Privilege of Partial Perspective', pp. 183–201 in Donna Haraway (ed.), *Simians, Cyborgs and Women: The Reinvention of Nature*, London: Free Association Books.

Haraway, Donna J. (1997), *Modest_Witness@Second_Millenium.Female_Man© _Meets_OncoMouse™ Feminism and Technoscience*, New York and London: Routledge.

Haraway, Donna J. (2003), *The Companion Species Manifesto: Dogs, People, and Significant Otherness*, Chicago: Prickly Paradigm Press.

Harvey, David (1993), 'Class Relations, Social Justice and the Politics of Difference', pp. 41–66 in Michael Keith and Steve Pile (eds), *Place and the Politics of Identity*, London and New York: Routledge.

Hesse, Mary B. (1963), *Models and Analogies in Science*, London: Sheed & Ward.

Hesse, Mary B. (1974), *The Structure of Scientific Inference*, London: Macmillan.

Hine, Christine (2000), *Virtual Ethnography*, Thousand Oaks, CA, London and New Delhi: Sage.

Hirschauer, Stefan (1998), 'Performing Sexes and Genders in Medical Practices', pp. 13–37 in Marc Berg and Annemarie Mol (eds), *Differences in Medicine: Unravelling Practices, Techniques and Bodies*, Durham, NC: Duke University Press.

Hirschauer, Stefan, and Annemarie Mol (1995), 'Shifting Sexes, Moving Stories: Feminist/Constructivist Dialogues', *Science, Technology and Human Values*, 20: 368–385.

Hockney, David (2001), *Secret Knowledge: Rediscovering the Lost Techniques of the Old Masters*, London: Thames & Hudson.

Ingold, Tim (2000), *The Perception of the Environment: Essays in Livelihood, Dwelling and Skill*, London and New York: Routledge.

Jameson, Frederic (1991), *Postmodernism, or, the Cultural Logic of Late Capitalism*, London: Verso.

Kerle, Anne (1995), *Uluru, Kata Tjuta & Watarrka: Ayers Rock, the Olgas & Kings Canyon, Northern Territory*, Sydney: University of New South Wales Press.

Klinenberg, Eric (2002), *Heat Wave: A Social Autopsy of Disaster in Chicago*, Chicago and London: Chicago University Press.

Knorr Cetina, Karin D. (1981), *The Manufacture of Knowledge: An Essay on the Constructivist and Contextual Nature of Science*, Oxford: Pergamon Press.

Knorr Cetina, Karin D. (1999), *Epistemic Cultures: How the Sciences Make Knowledge*, Cambridge, MA and London: Harvard University Press.

Kuhn, Thomas S. (1970), *The Structure of Scientific Revolutions*, Chicago: Chicago University Press.

Lacqueur, Thomas (1990), *Making Sex: Body and Gender from the Greeks to Freud*, Cambridge, MA: Harvard University Press.

Lakatos, Imre, and A. Musgrave (eds) (1970), *Criticism and the Growth of Knowledge*, Cambridge: Cambridge University Press.

Latour, Bruno (1987), *Science in Action: How to Follow Scientists and Engineers Through Society*, Milton Keynes: Open University Press.

Latour, Bruno (1988), *Irréductions*, published with *The Pasteurisation of France*, Cambridge, MA: Harvard University Press.

Latour, Bruno (1990), 'Drawing Things Together', pp. 19–68 in Michael Lynch and Steve Woolgar (eds), *Representation in Scientific Practice*, Cambridge, MA: MIT Press.

Latour, Bruno (1993), *We Have Never Been Modern*, Brighton: Harvester Wheatsheaf.

Latour, Bruno (1996), *Aramis, or the Love of Technology*, Cambridge, MA: MIT Press.

Latour, Bruno (1997), 'Foreword: Stengers's Shibboleth', pp. vii–xx in Isabelle Stengers (ed.), *Power and Invention: Situating Science*, Minneapolis and London: University of Minnesota Press.

Latour, Bruno (1998), *Pandora's Hope: Essays on the Reality of Science Studies*, Cambridge, MA: Harvard University Press.

Latour, Bruno, and Steve Woolgar (1986), *Laboratory Life: The Construction of Scientific Facts*, Second Edition, Princeton, NJ: Princeton University Press.

Law, John (1994), *Organizing Modernity*, Oxford: Blackwell.

Law, John (1998), 'After Metanarrative: On Knowing in Tension', pp. 88–108 in Robert Chia (ed.), *Into the Realm of Organisation: Essays for Robert Cooper*, London: Routledge.

Law, John (2000), 'Transitivities', *Society and Space*, 18: 133–148.

Law, John (2002a), *Aircraft Stories: Decentering the Object in Technoscience*, Durham, NC: Duke University Press.

Law, John (2002b), 'Economics as Interference', pp. 21–38 in Paul du Gay and Michael Pryke (eds), *Cultural Economy*, London: Sage.

Law, John (2002c), 'Objects and Spaces', *Theory, Culture and Society*, 19: 91–105.

Law, John, and Ruth Benschop (1997), 'Resisting Pictures: Representation, Distribution and Ontological Politics', pp. 158–182 in Kevin Hetherington and Rolland Munro (eds), *Ideas of Difference: Social Spaces and the Labour of Division*, Sociological Review Monograph, Oxford: Blackwell.

Law, John, and Annemarie Mol (1995), 'Notes on Materiality and Sociality', *Sociological Review*, 43: 274–294.

Law, John, and Annemarie Mol (1998), 'On Metrics and Fluids: Notes on Otherness', pp. 20–38 in Robert Chia (ed.), *Organised Worlds: Explorations in Technology, Organisation and Modernity*, London: Routledge.

Law, John, and Annemarie Mol (2001), 'Situating Technoscience: An Inquiry into Spatialities', *Society and Space*, 19: 609–621.

Law, John, and Annemarie Mol (eds) (2002), *Complexities: Social Studies of Knowledge Practices*, Durham, NC: Duke University Press.

Law, John, and Vicky Singleton (2003), 'Allegory and its Others', pp. 225–254 in Davide Nicolini, Silvia Gherardi, and Dvora Yanow (eds), *Knowing in Organizations: a Practice Based Approach*, New York: M.E. Sharpe.

Law, John, and Vicky Singleton (forthcoming), 'Object Lessons', *Organization*.

Law, John, and John Urry (2004), 'Enacting the Social', *Economy and Society*.

Lawson, Hilary (2001), *Closure: A Story of Everything*, London and New York: Routledge.

Layton, Robert (1989), *Uluru: An Aboriginal History of Ayers Rock*, Canberra: Aboriginal Studies Press.

London Yearly Meeting of the Religious Society of Friends (1960), *Christian Faith and Practice in the Experience of the Society of Friends*, London: Religious Society of Friends.

Lynch, Michael (1985), *Art and Artifact in Laboratory Science: A Study of Shop Work and Shop Talk in a Research Laboratory*, London: Routledge & Kegan Paul.

Mannheim, Karl (1953), 'Conservative Thought', pp. 74–164 in Karl Mannheim (ed.), *Essays on Sociology and Social Psychology*, London: Routledge & Kegan Paul.

Merton, Robert K. (1973a), 'The Normative Structure of Science', pp. 267–278 in Norman W. Storer (ed.), *The Sociology of Science*, Chicago: Chicago University Press.

Merton, Robert K. (1973b), 'Science and the Social Order', pp. 254–266 in Norman W. Storer (ed.), *The Sociology of Science*, Chicago: Chicago University Press.

Mol, Annemarie (1999), 'Ontological Politics: A Word and Some Questions', pp. 74–89 in John Law and John Hassard (eds), *Actor Network Theory and After*, Oxford and Keele: Blackwell and the Sociological Review.

Mol, Annemarie (2000), 'Pathology and the Clinic: An Ethnographic Presentation of Two Atheroscleroses', in Margaret Lock, Allan Young and Alberto Cambrosio (eds), *Living and Working with the New Medical Technologies*, Cambridge: Cambridge University Press.

Mol, Annemarie (2002), *The Body Multiple: Ontology in Medical Practice*, Durham, NC and London: Duke University Press.

Mol, Annemarie, and Marc Berg (1994), 'Principles and Practices of Medicine: The Coexistence of Various Anaemias', *Culture, Medicine and Psychiatry*, 18: 247–265.

Mol, Annemarie, and John Law (1994), 'Regions, Networks and Fluids: Anaemia and Social Topology', *Social Studies of Science*, 24: 641–671.

Moreira, Tiago (2000), 'Translation, Difference and Ontological Fluidity: Cerebral Angiography and Neurosurgical Practice', *Social Studies of Science*, 30: 421–446.

Moreira, Tiago (2001a), 'Incisions: A Study of Surgical Trajectories', PhD, Lancaster University.

Moreira, Tiago (2001b), 'Involvement and Constraint in a Surgical Consultation Room', *Bulletin Suisse de Linguistique Appliquée*, 13–32.

Moser, Ingunn (2000), 'Against Normalisation: Subverting Norms of Ability and Disability', *Science as Culture*, 9: 201–240.

Moser, Ingunn (2003), 'Living After Traffic Accidents: On the Ordering of Disabled Bodies', PhD, University of Oslo.

Moser, Ingunn, and John Law (1998), 'Materiality, Textuality, Subjectivity: Notes on Desire, Complexity and Inclusion', *Concepts and Transformation: International Journal of Action Research and Organizational Renewal*, 3: 207–227.

Moser, Ingunn, and John Law (1999), 'Good Passages, Bad Passages', pp. 196–219 in John Law and John Hassard (eds), *Actor Network and After*, Oxford and Keele: Blackwell and the Sociological Review.

Moser, Ingunn, and John Law (2003), '"Making Voices": New Media Technologies, Disabilities, and Articulation', pp. 491–420 in Gunnar Liestøl, Andrew Morrison and Terje Rasmussen (eds), *Digital Media Revisited: Theoretical and Conceptual Innovation in Digital Domains*, Cambridge, MA and London: MIT Press.

Nettleton, Sarah (1995), *The Sociology of Health and Illness*, Cambridge: Polity.

Osborne, Thomas, and Nikolas Rose (1999), 'Do the Social Sciences Create Phenomena? The Example of Public Opinion Research', *British Journal of Sociology*, 50: 367–396.

Perrow, Charles (1984), *Normal Accidents: Living with High Risk Technologies*, New York: Basic Books.

Phillips, Derek L. (1973), *Abandoning Method: Sociological Studies in Methodology*, San Francisco: Jossey-Bass.

Pickering, Andrew (1995), *The Mangle of Practice: Time, Agency and Science*, Chicago and London: University of Chicago Press.

Pinch, Trevor J. (1980), 'Theoreticians and the Production of Experimental Anomaly; the Case of Solar Neutrinos', in Karin D. Knorr, Roger Krohn and Richard D. Whitley (eds), *The Social Processes of Scientific Investigation, Sociology of the Sciences*, Vol. 4, Dordrecht, Boston and London: Reidel.

Pinch, Trevor J. (1981), 'The Sun Set: The Presentation of Certainty in Scientific Life', *Social Studies of Science*, 11: 131–158.

Pinch, Trevor J. (1985), 'Towards an Analysis of Scientific Observation: The Externality and Evidential Significance of Observational Reports in Science', *Social Studies of Science*, 15: 3–36.

Polanyi, Michael (1958), *Personal Knowledge: Towards a Post-Critical Philosophy*, London: Routledge & Kegan Paul.

Popper, Karl R. (1959), *The Logic of Scientific Discovery*, London: Hutchinson.

Porter, Theodore M. (1995), *Trust in Numbers: The Pursuit of Objectivity in Science and Public Life*, Princeton, NJ: Princeton University Press.

Power, Michael (1997), *The Audit Society: Rituals of Verification*, Oxford: Oxford University Press.

Ravetz, Jerome R. (1973), *Scientific Knowledge and its Social Problems*, Harmondsworth: Penguin.

Rip, Arie (1986), 'Mobilizing Resources Through Texts', in Michel Callon, John Law and Arie Rip (eds), *Mapping the Dynamics of Science and Technology: Sociology of Science in the Real World*, Basingstoke: Macmillan.

Robert, Le (1974), *Dictionnaire Alphabétique et Analogique de la Langue Française*, Paris: Le Robert.

Rock, Paul (1979), *The Making of Symbolic Interactionism*, London: Macmillan.

Rose, Gillian (2001), *Visual Methodologies*, London, Thousand Oaks, CA and New Delhi: Sage.

Rose, Nikolas (1999), *Powers of Freedom: Reframing Political Thought*, Cambridge: Cambridge University Press.

Rotman, Brian (1987), *Signifying Nothing: The Semiotics of Zero*, Stanford: Stanford University Press.

Said, Edward W. (1991), *Orientalism: Western Conceptions of the Orient*, London: Penguin.

Saussure, de (1960), *Course in General Linguistics*, London: Peter Owen.

Sayer, Andrew (2000), *Realism and Social Science*, London, Thousand Oaks, CA and New Delhi: Sage.

Scarry, Elaine (1985), *The Body in Pain: The Making and Unmaking of the World*, New York and Oxford: Oxford University Press.

Serres, Michel (1980), *Le Passage du Nord-Ouest*, Hermes V, Paris: Les Éditions de Minuit.

Shapin, Steven (1984), 'Pump and Circumstance: Robert Boyle's Literary Technology', *Social Studies of Science*, 14: 481–520.

Shapin, Steven (1989), 'The Invisible Technician', *American Scientist*, 77: 554–563.

Shapin, Steven (1994), *A Social History of Truth: Civility and Science in Seventeenth-Century England*, Chicago: Chicago University Press.

Shapin, Steven, and Simon Schaffer (1985), *Leviathan and the Air Pump: Hobbes, Boyle and the Experimental Life*, Princeton, NJ: Princeton University Press.

Sharp, Nonie (1996), *No Ordinary Judgement: Mabo, the Murray Islanders' Land Case*, Canberra: Aboriginal Studies Press.

Sherlock, Sheila (1989), *Diseases of the Liver and Biliary System*, 8th edn, Oxford, London, Edinburgh, Boston and Melbourne: Blackwell.

Singleton, Vicky (1996), 'Feminism, Sociology of Scientific Knowledge and Postmodernism: Politics, Theory and Me', *Social Studies of Science*, 26: 445–468.

Singleton, Vicky (1998), 'Stabilizing Instabilities: The Role of the Laboratory in the

United Kingdom Cervical Screening Programme', pp. 86–104 in Marc Berg and Annemarie Mol (eds), *Differences in Medicine: Unravelling Practices, Techniques and Bodies*, Durham, NC: Duke University Press.

Singleton, Vicky, and Mike Michael (1993), 'Actor-networks and Ambivalence: General Practitioners in the UK Cervical Screening Programme', *Social Studies of Science*, 23: 227–264.

Starhawk (1989), *The Spiral Dance: A Rebirth of the Ancient Religion of the Great Goddess*, New York and San Francisco: Harper.

Stengers, Isabelle (1997), *Power and Invention: Situating Science*, Minneapolis and London: University of Minnesota Press.

Strathern, Marilyn (1991), *Partial Connections*, Savage Maryland: Rowman & Littlefield.

Sutton, Peter (ed.) (1989), *Dreamings: the Art of Aboriginal Australia*, Ringwood, Victoria and London: Viking.

Thompson, E.P. (1967), 'Time, Work-Discipline, and Industrial Capitalism', *Past and Present*, 38: 56–96.

Thrift, Nigel (1996), *Spatial Formations*, London, Thousand Oaks, CA and New Delhi: Sage.

Thrift, Nigel (2000), 'Afterwords', *Society and Space*, 18: 213–255.

Traweek, Sharon (1988), *Beamtimes and Lifetimes: The World of High Energy Physics*, Cambridge, MA: Harvard University Press.

Traweek, Sharon (1999), 'Pilgrim's Progress: Male Tales Told During a Life in Physics', pp. 525–542 in Mario Biagioli (ed.), *The Science Studies Reader*, New York and London: Routledge.

Turkle, Sherry (1996), *Life on the Screen: Identity in the Age of the Internet*, London: Weidenfeld & Nicolson.

Turnbull, David (1993), *Maps are Territories, Science is an Atlas*, Chicago: Chicago University Press.

Turnbull, David (1996), 'Cartography and Science in Early Modern Europe: Mapping the Construction of Knowledge Spaces', *Imago Mundi*, 48: 5–24.

Turnbull, David (2000), *Masons, Tricksters and Cartographers: Comparative Studies in the Sociology of Scientific and Indigenous Knowledge*, Amsterdam: Harwood Academic Publishers.

Turnbull, David (forthcoming), 'Locating, Negotiating, and Crossing Boundaries: A Western Desert Land Claim, The Tordesillas Line and The West Australian Border', *Society and Space*.

Urry, John (2000), 'Mobile Sociology', *British Journal of Sociology*, 51: 185–203.

Verran, Helen (1998), 'Re Imagining Land Ownership in Australia', *Postcolonial Studies*, 1: 237–254.

Verran, Helen (2001), *Science and an African Logic*, Chicago and London: Chicago University Press.

Verran, Helen (2002), 'Transferring Strategies of Land Management: Indigenous Land Owners and Environmental Scientists', pp. 155–181 in Marianne de Laet (ed.), *Research in Science and Technology Studies*, Vol. 13: *Knowledge and Technology Transfer*, New York: JAI Press.

Verran, Helen (forthcoming), *Science and the Dreaming: Expertise in a Complex World*, Chicago and London: Chicago University Press.

Watson-Verran, Helen, and David Turnbull (1995), 'Science and other Indigenous Knowledge Systems', pp. 115–139 in Sheila Jasanoff, Gerard E. Markle, James C.

Petersen and Trevor Pinch (eds), *Handbook of Science and Technology Studies*, Thousand Oaks, CA: Sage.

Williams, Raymond (1988), *Keywords: A Vocabulary of Culture and Society*, London: Fontana Press.

Wilson, Harold (1971), *The Labour Government, 1964–1970: A Personal Record*, London: Weidenfeld & Nicolson and Michael Joseph.

Wittgenstein, Ludwig (1953), *Philosophical Investigations*, Oxford: Blackwell.

Wynne, Brian (1996), 'May the Sheep Safely Graze? A Reflexive View of the Expert–Lay Knowledge Divide', pp. 44–83 in Scott Lash, Bronislaw Szerszynski and Brian Wynne (eds), *Risk, Environment and Modernity: Towards a New Ecology*, London and Beverly Hills, CA: Sage.

Index

eBooks – at www.eBookstore.tandf.co.uk

A library at your fingertips!

eBooks are electronic versions of printed books. You can store them on your PC/laptop or browse them online.

They have advantages for anyone needing rapid access to a wide variety of published, copyright information.

eBooks can help your research by enabling you to bookmark chapters, annotate text and use instant searches to find specific words or phrases. Several eBook files would fit on even a small laptop or PDA.

NEW: Save money by eSubscribing: cheap, online access to any eBook for as long as you need it.

Annual subscription packages

We now offer special low-cost bulk subscriptions to packages of eBooks in certain subject areas. These are available to libraries or to individuals.

For more information please contact webmaster.ebooks@tandf.co.uk

We're continually developing the eBook concept, so keep up to date by visiting the website.

www.eBookstore.tandf.co.uk